Biochemistry Research Trends

Biochemistry Research Trends

Mineral Water: From Basic Research to Clinical Applications
Maria João Martins, PhD (Editor)
2022. ISBN: 978-1-68507-458-6 (Hardcover)
2022. ISBN: 978-1-68507-541-5 (eBook)

Terpenes and Terpenoids: Sources, Applications and Biological Significance
Charles A. Davies (Editor)
2022. ISBN: 978-1-68507-559-0 (Hardcover)
2022. ISBN: 978-1-68507-595-8 (eBook)

Circadian Rhythms and Their Importance
Rajeshwar P. Sinha, PhD (Editor)
2022. ISBN: 978-1-68507-547-7 (Hardcover)
2022. ISBN: 978-1-68507-585-9 (eBook)

A Biochemical View of Antioxidants
David Aebisher PhD, DSc, Dorota Bartusik-Aebisher, PhD (Editors)
2021. ISBN: 978-1-68507-151-6 (Hardcover)
2021. ISBN: 978-1-68507-295-7 (eBook)

Volatile Oils: Production, Composition and Uses
Sunita Singh, PhD (Editor)
2021. ISBN: 978-1-68507-186-8 (Hardcover)
2021. ISBN: 978-1-68507-241-4 (eBook)

An Introduction to Drug Carriers
Mohammad Ashrafuzzaman, D.Sc. (Editor)
2021. ISBN: 978-1-68507-148-6 (Hardcover)
2021. ISBN: 978-1-68507-157-8 (eBook)

More information about this series can be found at
https://novapublishers.com/product-category/series/biochemistry-research-trends/

**David Aebisher and
Dorota Bartusik-Aebisher**
Editors

The Biochemical Guide to Proteins

Copyright © 2023 by Nova Science Publishers, Inc.

https://doi.org/10.52305/UIEP1882

All rights reserved. No part of this book may be reproduced, stored in a retrieval system or transmitted in any form or by any means: electronic, electrostatic, magnetic, tape, mechanical photocopying, recording or otherwise without the written permission of the Publisher.

We have partnered with Copyright Clearance Center to make it easy for you to obtain permissions to reuse content from this publication. Simply navigate to this publication's page on Nova's website and locate the "Get Permission" button below the title description. This button is linked directly to the title's permission page on copyright.com. Alternatively, you can visit copyright.com and search by title, ISBN, or ISSN.

For further questions about using the service on copyright.com, please contact:
Copyright Clearance Center
Phone: +1-(978) 750-8400   Fax: +1-(978) 750-4470   E-mail: info@copyright.com

NOTICE TO THE READER

The Publisher has taken reasonable care in the preparation of this book, but makes no expressed or implied warranty of any kind and assumes no responsibility for any errors or omissions. No liability is assumed for incidental or consequential damages in connection with or arising out of information contained in this book. The Publisher shall not be liable for any special, consequential, or exemplary damages resulting, in whole or in part, from the readers' use of, or reliance upon, this material. Any parts of this book based on government reports are so indicated and copyright is claimed for those parts to the extent applicable to compilations of such works.

Independent verification should be sought for any data, advice or recommendations contained in this book. In addition, no responsibility is assumed by the Publisher for any injury and/or damage to persons or property arising from any methods, products, instructions, ideas or otherwise contained in this publication.

This publication is designed to provide accurate and authoritative information with regard to the subject matter covered herein. It is sold with the clear understanding that the Publisher is not engaged in rendering legal or any other professional services. If legal or any other expert assistance is required, the services of a competent person should be sought. FROM A DECLARATION OF PARTICIPANTS JOINTLY ADOPTED BY A COMMITTEE OF THE AMERICAN BAR ASSOCIATION AND A COMMITTEE OF PUBLISHERS.

Additional color graphics may be available in the e-book version of this book.

Library of Congress Cataloging-in-Publication Data

ISBN: 979-8-88697-493-5

Published by Nova Science Publishers, Inc. † New York

Contents

Preface ix

Chapter 1 **Tubulin** 1
Damian Bezara, Dorota Bartusik-Aebisher
and David Aebisher

Chapter 2 **Transforming Growth Factors (TGFs)** 7
Piotr Czerniak, Dorota Bartusik-Aebisher
and David Aebisher

Chapter 3 **Elastin** 11
Weronika Bargiel, Dorota Bartusik-Aebisher
and David Aebisher

Chapter 4 **Collagen** 17
Julia Buszek, Dorota Bartusik-Aebisher
and David Aebisher

Chapter 5 **Myosin** 23
Dominik Jaklik, Dorota Bartusik-Aebisher
and David Aebisher

Chapter 6 **Fibronectin** 27
Jacek Mazanek, Dorota Bartusik-Aebisher
and David Aebisher

Chapter 7 **Colony-Stimulating Factors (CSFs)** 33
Adrian Groele, Dorota Bartusik-Aebisher
and David Aebisher

Chapter 8 **Histones** 39
Agnieszka Zaleszczyk, Dorota Bartusik-Aebisher
and David Aebisher

Chapter 9	**Protamines** 43	
	Agnieszka Gancarz, Dorota Bartusik-Aebisher and David Aebisher	
Chapter 10	**Enkephalins as Peptides Widely Distributed in the Body** 49	
	Natalia Guzik, Dorota Bartusik-Aebisher and David Aebisher	
Chapter 11	**Platelet-Derived Growth Factor (PDGF)** 55	
	Monika Błądek, Dorota Bartusik-Aebisher and David Aebisher	
Chapter 12	**Immunoglobulins** 59	
	Maria Dycha, Dorota Bartusik-Aebisher and David Aebisher	
Chapter 13	**SGLT1 in Head and Neck Cancers** 65	
	Lidia Bieniasz, Dorota Bartusik-Aebisher and David Aebisher	
Chapter 14	**Major Histocompatibility Antigens** 69	
	Klaudia Fikas, Dorota Bartusik-Aebisher and David Aebisher	
Chapter 15	**Vascular Endothelial Growth Factor (VEGF)** 75	
	Eliza Gryboś, Dorota Bartusik-Aebisher and David Aebisher	
Chapter 16	**T Cell Receptor** 81	
	Paulina Nowak, Dorota Bartusik-Aebisher and David Aebisher	
Chapter 17	**Serum Albumin** 87	
	Kamil Jugo, Dorota Bartusik-Aebisher and David Aebisher	
Chapter 18	**Ferritin** 93	
	Kamil Chwaliszewski, Dorota Bartusik-Aebisher and David Aebisher	

Chapter 19	**Keratin**..99	
	Adrianna Antoszewska, Dorota Bartusik-Aebisher and David Aebisher	
Chapter 20	**GroEL**...105	
	Karolina Drygała, Dorota Bartusik-Aebisher and David Aebisher	
Chapter 21	**Fibroblast Growth Factor (FGF)**..................................111	
	Zuzanna Wielgosz, Dorota Bartusik-Aebisher and David Aebisher	
Chapter 22	**Hemoglobin**...117	
	Szymon Płaneta, Dorota Bartusik-Aebisher and David Aebisher	
Chapter 23	**p53**..123	
	Martyna Lipian, Dorota Bartusik-Aebisher and David Aebisher	
Chapter 24	**Chitinase**...129	
	Mateusz Pomianek, Dorota Bartusik-Aebisher and David Aebisher	
Chapter 25	**B-raf**...135	
	Julia Michalik, Dorota Bartusik-Aebisher and David Aebisher	
Chapter 26	**Rhodopsin**..141	
	Karolina Miś, Dorota Bartusik-Aebisher and David Aebisher	
Chapter 27	**Estrogen Receptors**...147	
	Julia Kudła, Dorota Bartusik-Aebisher and David Aebisher	
Chapter 28	**Ceruloplasmin**..153	
	Natalia Magierło, Dorota Bartusik-Aebisher and David Aebisher	
Chapter 29	**Prostate-Specific Antigen (PSA)**...................................159	
	Halszka Wajdowicz, Dorota Bartusik-Aebisher and David Aebisher	

Chapter 30	HER2	165
	Kacper Rogóż, Dorota Bartusik-Aebisher and David Aebisher	
Index		171
About the Editors		175

Preface

This book aims to provide scientific information about proteins. 30 chapters have been written that describe the characteristics of proteins and their influence on human physiology. The authors presented the structure of individual proteins and their role in particular diseases. Proteins are among the most important nutrients responsible for the proper functioning of the body. Protein is one of the basic building blocks of our tissues, plays a role in transport, regulating biochemical processes and in reactions initiated by the immune system.

Chapter 1

Tubulin

Damian Bezara, Dorota Bartusik-Aebisher[*] and David Aebisher
Medical College of The University of Rzeszów, Rzeszów, Poland

Abstract

Tubulin is a universal and key protein for the functioning of cells, especially in eukaryotes. Tubulins belonging to the alpha, beta and gamma groups belong to the universal tubulins involved in key processes of the cell cycle. Non-universal tubulins are not present in all organisms, but are characteristic only of some of them, for example zeta-tubulin absent from Homo sapiens sapiens, where its role is played by delta-tubulin. A, β, γ tubulins are modified by attaching chemical groups to the side chains of their amino acids and binding them to other proteins. These modifications change the properties of the macrostructure they create, thus influencing its lifespan or interactions with other proteins or non-protein components of the cytosol.

Keywords: tubulin, microtubules, proteins, glutamylation, tyrosination, microtubule organization site (MTOC)

Tubulin is a globular protein with different variants that occurs in the cell in the form of microtubules. The microtubule is composed of two heterodimers (Figure 1), i.e., two connected alpha and beta tubulin chains, arranged concentrically, so that there are 13 heterodimers for every 360o. Concentric systems of heterodimers combine to form a long thread-like structure, along

---

[*] Corresponding Author's Email: dbartusikaebisher@ur.edu.pl.

In: The Biochemical Guide to Proteins
Editors: David Aebisher and Dorota Bartusik-Aebisher
ISBN: 979-8-88697-493-5
© 2023 Nova Science Publishers, Inc.

the long axis of which there is an alternating arrangement of alpha and beta tubulin subunits. Microtubules have 2 ends "+" and "-" (Gadadhar et al. 2017). The end marked with a positive sign means the dynamic addition of further tubulins to the thread, and the negative end – depolymerization of microtubules. The properties of microtubules (mainly the half-life and its function) depend both on the structure of individual types of tubulin resulting from the primary structure and post-translational modification, as well as on accompanying proteins, e.g., the microtubule stabilizing protein Tau. Another example of tubulin is gamma fibrillar, similar to eukaryotic beta tubulin and bacterial tubulin. Its function is to create tetramers, which are the "seed" for the formation of microtubules in the centrosomes, and to stabilize the "+" pole of the microtubules. Microtubules take part in vital processes for the cell, which include: karyokinesis, intracellular vesicular transport, or the formation of mobile cilia and flagella. In addition to the previously mentioned, there are also δ, ε, ζ and η tubulin subunits (Gadadhar et al. 2017). The aim of this article is to present the structure of tubulin, taking into account its various variants, with reference to the functions of the microtubule in relation to the differences in its components.

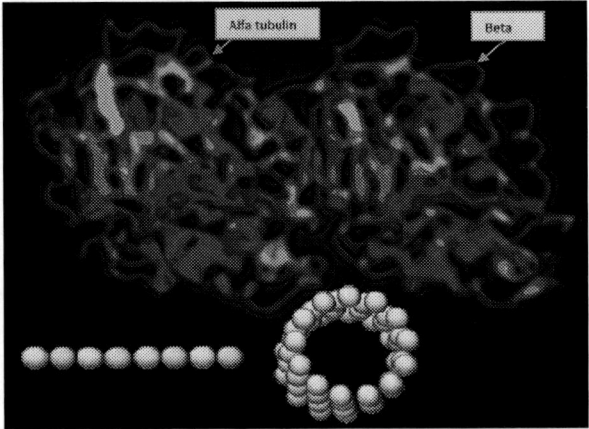

Figure 1. Heterodimer of the alpha and beta tubulin subunits A heterodimer composed of alpha (top left) and beta tubulin (top right) subunits is part of the microtubule. Heterodimers are arranged with a long axis parallel to the long axis of the microtubule, the arrangement of alpha (gray) and beta (gold) subunits are shown in two diagrams, on the lower left the heterodimers arrangement with respect to the long axis of the microtubule, on the right a cross section of the microtubule. The alpha-helices are marked in red, beta-sheets in green, and the remaining sections of the polypeptide in blue [Own elaboration].

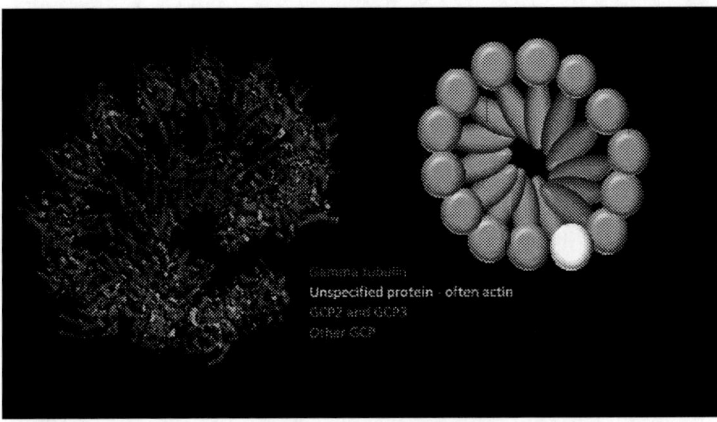

Figure 2. Gamma tubulin complex, The gamma tubulin complex usually consists of 28 distinct associated structures, the outer half-ring in the illustration corresponds to a number of 13 gamma-tubulin subunits. 14 of them are an undefined protein – often actin. The proteins forming the inner half-ring belong to the GCP series of proteins, GCP2 and GCP3 are the dominant ones [Own elaboration].

The diversity of tubulin subunits is caused by a large genetic diversity manifested by the presence of genes for several types of subunits in the genetic code of the eukaryote, as well as a large variety of post-translational modifications. Humans have eight α- and nine β-tubulin genes. Post-translational modifications are related to changes in the structure of individual subunits, which affect both the physical and chemical properties of the structures they build. The difference in their properties results, for example, from a different affinity for the proteins accompanying microtubules, which means that a given type of microtubule has a different behavior and durability, which is important, e.g., in the case of the formation of a division spindle – a labile structure, as well as stable centrosomes (Janke et al. 2020). These modifications include acetylation of α and β-tubulin, phosphorylation of α and β-tubulin, polyamination, detyrosination/ tyrosination (removal and addition of genetically encoded C-terminal tyrosine on α-tubulin), Δ2-tubulin (removal of last and penultimate glutamates in α- tubulin), glutamylation (the reversible addition of branched chain glutamates of variable length to α or β-tubulin) and glycylation (the addition of branched chain glycine of variable length to α or β-tubulin). Glutamylation and tyrosination of α and β-tubulins increase their ability to recruit dynein – a motor protein that participate in transport pathways in guanylated microtubules, and are responsible for flagella movement in tyrosinated microtubules (Janke et al. 2020).

Y tubulin forms the γ-TurRC complex consisting of 13 subunits of this protein and one subunit of indefinite protein, and a number of GCP proteins (GCP1 - GCP6) associating with them into a spiral-shaped structure (Figure 2). This complex acts as a sporulating region in the cell for the formation of microtubules, but the mechanism is not yet fully understood. It probably relies on facilitating the flanking of short alpha dimers, beta tubulin, through their interaction with gamma tubulins. This is important because the structure of a short microtubule up to a certain length is kinetically unfavorable, which translates into the inability to quickly polymerize. Even more difficult is the mechanism of formation of the γ-TurRC complexes themselves, mainly because not all of them contain all the standard components. Additional proteins include Grip71 mediating the interaction between the proteins of the γ-TurRC complex, and MZT1, MZT2 – regulating the interaction between γ-TurRC and the microtubule organization site (MTOC) in the cell, NME7 – a kinase that increases the activity of nucleation. The assembly of γ-TurRC itself occurs differently in the cytosol within and outside the MTOC; the cytosolic assembly does not involve MTOC and related factors but requires the presence of GCP 4, 5, 6, while the MTOC-dependent association is dependent on the presence of specific Mto1 and Mto2 complexes (Lopes et al. 2020).

Delta tubulin is a protein involved in the formation of a distinctive centriole pattern, which is described as 9 concentric triplicate microtubules with one pair in the center. The delta tubulin deficiency or abnormal structure is responsible for the formation of centrioles with nine pairs rather than three concentrically arranged microtubules. Epsilon tubulin was discovered in further studies on the function of delta tubulin, egg labeling with antibodies allowed for the detection of this variety in the parent centriole; probably in the centrosome it does not occur in daughter centrioles, and its function is related to the duplication of centrioles (Lin et al. 2020). Zeta tubulin is a protein that forms complexes with delta and epsilon tubulin called ZED. It is postulated that zeta tubulin is involved in the interaction between the cytoskeleton and centriole, which enables the cell to determine its polarity during division. Particularly high content of zeta tubulin characterizes multi-ciliated cells, moreover, it is a non-human protein, in the body of which delta tubulin performs some of its functions. Eta tubulin is the least understood of all tubulin subunits described herein. Its function is virtually unknown, but a genetic comparison with other variants of this protein led to the assumption that eta-tubulin contributes to the binding of gamma tubulin to the site of formation of the basal bodies. Tubulins belonging to this group are characterized by the lack of polymerization capacity and occurring only in the complex they form with centriole, except for protist cells, animals and algae, in which they also form a basal body (Roll-Mecak et al. 2020).

Tubulin is a universal and key protein for the functioning of cells, especially in eukaryotes. New varieties are constantly being discovered and the multitude of functions fulfilled by these proteins is surprising. Tubulins belonging to the alpha, beta and gamma groups belong to the universal tubulins involved in key processes of the cell cycle (gamma tubulin participates in the initiation of microtubule formation, alpha and beta tubulins form a karyokinetic spindle), movement (cilia, flagella made of alpha and beta tubulin) and their specific functions (intracellular transport pathways). In some species it is possible to include epsilon tubulin in the above group due to its importance in the duplication of centrioles. Non-universal tubulins are not present in all organisms, but are characteristic only of some of them, for example zeta-tubulin absent from Homo sapiens sapiens, where its role is played by delta-tubulin. Tubulins from groups δ, ε, ζ, η do not polymerize in microtubules or helical-shaped systems characteristic for γ-tubuulin, but form associations with centrioles. Their functions can be taken over by the remaining tubulins or other groups of proteins. A, β, γ tubulins are modified by attaching chemical groups to the side chains of their amino acids and binding them to other proteins, e.g., Tau with α and β tubulins, and GCP with γ-tubulins. These modifications change the properties of the macrostructure they create, thus influencing its lifespan or interactions with other proteins or non-protein components of the cytosol.

References

Gadadhar, S., Bodakuntla, S., Natarajan, K., and Janke, C. (2017). The tubulin code at a glance. *J Cell Sci.*, 130(8), 1347-1353.

Janke, C., and Magiera, M. M. (2020). The tubulin code and its role in controlling microtubule properties and functions. *Nat Rev Mol Cell Biol.*, 21(6), 307-326.

Lin, Z., Gasic, I., Chandrasekaran, V., Peters, N., Shao, S., Mitchison, T. J., and Hegde, R. S. (2020). TTC5 mediates autoregulation of tubulin via mRNA degradation. *Science.*, 367(6473), 100-104.

Lopes, D., and Maiato, H. (2020). The Tubulin Code in Mitosis and Cancer. *Cells.*, 9(11), 2356.

Roll-Mecak, A. (2020). The Tubulin Code in Microtubule Dynamics and Information Encoding. *Dev Cell.*, 54(1), 7-20.

Chapter 2

Transforming Growth Factors (TGFs)

Piotr Czerniak, Dorota Bartusik-Aebisher[*] and David Aebisher
Medical College of The University of Rzeszów, Rzeszów, Poland

Abstract

Transforming growth factors (TGFs) are polypeptide compounds. TGF's with relatively low molecular weight, are thermally stable acid inactivated by agents reducing disulfide bonds. Uncontrolled proliferation is a characteristic feature of cancer development. Proteins from the TGF-beta family are inhibitors of cell growth, so their malfunction favors the uncontrolled growth of neoplastic tissue. The results of the current research confirm the increase in mRNA and TGF-beta protein expression in cancers of the pancreas, colon, lung, bone and the brain. The huge role of TGFs is played by transforming growth factors in the human body. Continuation of research on TGF-beta and its relationship to the development of neoplastic diseases, may be promising in the context of cancer disease.

Keywords: transforming growth factors (TGFs), glucocorticosteroid, bone morphogenetic protein (BMP), DNA

Transforming growth factors are polypeptide compounds of relatively low molecular weight. TGFs regulate processes related to growth, division,

---

[*] Corresponding Author's Email: dbartusikaebisher@ur.edu.pl.

In: The Biochemical Guide to Proteins
Editors: David Aebisher and Dorota Bartusik-Aebisher
ISBN: 979-8-88697-493-5
© 2023 Nova Science Publishers, Inc.

maturation, mutual communication, differentiation and phenotypic transformation of many cells (both changed and unchanged). TGF alpha and TGF beta, fulfill different functions and interact with the EGF receptor in a different way. (epidermal growth factor), these differences allow the classification of TGF to alpha or beta (Guttmann-Gruber et al., 2021).

TGF alpha is single-chain peptide composed of 50-53 amino acids, which compete with EGF for binding to its receptor. TGF beta, is composed of two chains (each of them contains: 112 amino acids, 9 cysteine residues) (Figure 1), they attach to specific receptors on the cell membrane on which they interact, depending on the conditions, they may have an inhibitory effect. stimulating the growth of the same cells. TGF alpha and TGF beta can interact both antagonistically and synergistically (Nickel et al., 2018).

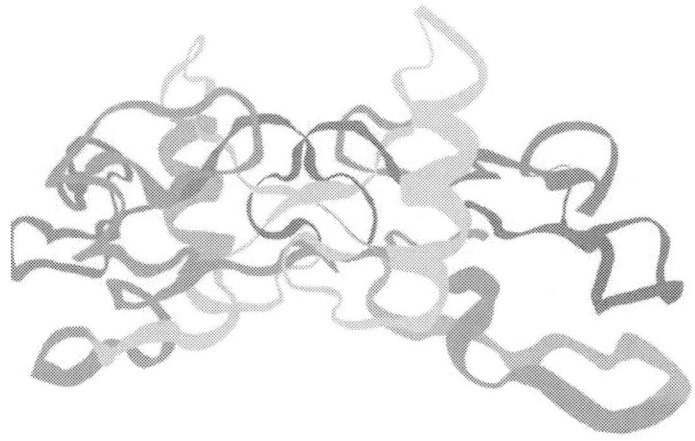

Figure 1. TGF-beta - connected polypeptide chains forming a three-dimensional form.

As a result of studies carried out on animal models, a relationship was found between the presence of TGF-beta and the speed of wound healing. The healing process includes many processes such as: migration and proliferation of cells, synthesis of extracellular substances or the formation of capillaries (angiogenesis), and TGF-beta is responsible for the regulation of many of these phenomena, so it accelerates the healing of damaged tissues, both superficial and and deep wounds. Transforming growth factor beta is also responsible for stimulating angiogenesis by local release of other growth factors, thanks to which it enables better blood flow and its reaching the sites of damage. Thanks to its properties to counteract the adverse effects of glucocorticosteroid drugs on the healing process, TGF-beta can be used in the treatment of difficult-to-heal wounds or chronic ulcers, e.g., in patients whose natural healing processes are reduced by therapy with steroids, radiation or other drugs showing such an effect (Qi et al., 2018).

Scientists studying the human genome discovered 28 genes encoding proteins from the TGF-beta family, about 40 different cytokines belong to it, for example, GDF (growth and activation factor) or BMP (bone morphogenetic protein). We can distinguish 5 isoforms of TGF-beta, of which only 3 occur in humans: TGF-β I, TGF-β II and TGF-β III. Each of the forms has a similar structure and shows a similar effect on cells (by acting on the same signal receptors). Speaking of TGF-beta, we cannot fail to mention the disturbances in its expression as a result of the action of ultraviolet radiation. Every living organism on earth is exposed to UVR, it can have both positive and negative effects on a living organism (Stocco et al., 2020).

The negative effects include: sunburn, pre-cancerous and neoplastic changes, and photoaging of the skin. UV rays can disrupt the functioning of some cytokines and their receptors on the cell surface. An increase in the expression of genes responsible for: TGF-beta type I and III was found, without reducing the expression of TGF-beta II. With increased exposure of the skin to the sun, the production of collagen decreases UV rays inhibit the expression of genes encoding TGF-beta, which participates in the synthesis of collagen. It is also worth paying attention to the participation of TGF-beta in renal parenchyma fibrosis, which is characterized by: an increase in the level of protein in the extracellular matrix, disturbance of cell-matrix regulation, reduced matrix degradation, infiltration of inflammatory cells and transformation of resident cells. The described situation is related to the wide range of functions performed by TGF-beta (cell proliferation and differentiation, regulation of immune reactions). Research confirms the direct effect of TGF-beta in renal glomerular diseases and the correlation between the increased expression of genes responsible for TGF-beta1 and areas of increased fibrosis, which is used to alleviate renal fibrosis with anti-TGF-beta therapy (Zhang et al., 2017).

The wide spectrum of functions fulfilled by TGF-beta showed a correlation between disturbances in its functioning and the occurrence of neoplasms. If cell malignancy is not inhibited by TGF-beta and increased expression of this factor occurs, it may increase the chances of tumor survival, as TGF has angiogenic and immunosuppressive effects, which facilitates tumor growth and protects it against the body's immune system. Specific genetic changes are observed in the signaling/inhibitory pathways involving

TGF-beta. These changes relate to defects in TGF-beta receptors, activation of genes related to TGF-beta and proteins that regulate the cell cycle. To sum up, we notice the huge role played by transforming growth factors in the human body, they fulfill a number of important functions, TGF-beta is responsible for the regulation of cell division, the result of irregularities in the functioning of this factor by e.g., genetic mutations may be the formation of neoplasms. Continuation of research on TGF-beta and its relationship to the development of neoplastic diseases is promising in the context of fighting these diseases.

References

Guttmann-Gruber C., Piñón Hofbauer J. Transforming growth factor-β messaging: #ContextMatters. *Br J Dermatol.* 2021 Apr;184(4):592-593.

Nickel J., Ten Dijke P., Mueller T. D. TGF-β family co-receptor function and signaling. *Acta Biochim Biophys Sin* (Shanghai). 2018 Jan 1;50(1):12-36.

Qi M., Zhou Q., Zeng W., Wu L., Zhao S., Chen W., Luo C., Shen M., Zhang J., Tang C. E. Growth factors in the pathogenesis of diabetic foot ulcers. *Front Biosci* (Landmark Ed). 2018 Jan 1;23:310-317.

Stocco E., Barbon S., Tortorella C., Macchi V., De Caro R, Porzionato A. Growth Factors in the Carotid Body-An Update. *Int J Mol Sci.* 2020 Oct 1;21(19):7267.

Zhang Y. E. Non-Smad Signaling Pathways of the TGF-β Family. *Cold Spring Harb Perspect Biol.* 2017 Feb 1;9(2):a022129.

Chapter 3

Elastin

Weronika Bargiel, Dorota Bartusik-Aebisher[*] and David Aebisher
Medical College of The University of Rzeszów, Rzeszów, Poland

Abstract

Elastin is a unique protein due to its limited synthesis time, long half-life, multi-stage assembly process and a high degree of cross-linking, which provides mechanical flexibility to large arteries. Despite increasing therapeutic opportunities to control diseases such as hypertension, obesity and diabetes, the elastic fiber damage that often accompanies these diseases is now irreparable, meaning that cardiovascular risks from arterial stiffness may persist even when other conditions are present. Aspects of the disease are controlled.

Elastin has also been shown to be associated with photoaging and the occurrence of SVAS congenital heart disease – supravalvular aortic stenosis.

Keywords: elastin, supravalvular aortic stenosis (SVAS), large latent complex (LLC), elastogenesis

Introduction

Elastin is a key protein in the extracellular matrix. It is the main component of elastic fibers which, for example, ensure the flexibility of large arteries

---

[*] Corresponding Author's Email: dbartusikaebisher@ur.edu.pl.

In: The Biochemical Guide to Proteins
Editors: David Aebisher and Dorota Bartusik-Aebisher
ISBN: 979-8-88697-493-5
© 2023 Nova Science Publishers, Inc.

necessary for the proper functioning of the circulatory system of vertebrates. It is also crucial for the flexibility of ligaments, tendons, skin and elastic cartilage. Elastin deficiency or various modifications of elastic fibers (disorganization, improper assembly, fragmentation and biochemical modifications) change the mechanical behavior of large arteries and affect cardiovascular mechanics. Elastic fibers undergo minimal rotation throughout their life and are very durable under various physiological stresses. For example, the elastic fibers in the wall of arteries allow them to undergo more than two billion cycles of stretching and relaxation to ease blood flow down the arterial tree with largely no mechanical damage (Wang et al., 2020).

The formation of elastic fibers, also known as elastogenesis, begins in the middle of pregnancy, reaches its maximum level shortly before birth, and ends during development in the postnatal life. The precursor of elastin is tropoelastin, which combines with other tropoelastin molecules through coacervation during the main phase of elastogenesis. Massively cross-linked tropoelastin systems (usually in combination with microfibers) contribute to the structural integrity and biomechanics of the tissue. Elastin sequences interact with many proteins found in microfiber and bind to elastogenic receptors on the cell surface.

Knowledge of the major steps involved in elastin formation has facilitated the construction of in vitro elastogenesis models, and has led to the identification of precise molecular regions critical to elastin-based protein interactions (Vindin et al., 2019). Elastin- functions in the body and biomedical significance (discussion).

Molecular Characterization

In mammals, the genome contains only one gene for tropoelastin, called ELN. The human ELN gene is a 45 kb segment encoded on chromosome 7 and has 34 exons interrupted by nearly 700 introns, the first exon being a signal peptide assigning the extracellular localization of the protein. The large number of introns suggests that genetic recombination may contribute to gene instability, leading to diseases such as SVAS.

SVAS

Supravalvular aortic stenosis (SVAS) is a congenital heart defect. As the name suggests, SVAS stands for a narrowing of the section of the aorta just above the aortic valve. SVAS accounts for 8% to 14% of all congenital aortic stenoses.

Elastin as a Signaling Molecule

Elastic fibers can be degraded and fragmented as a result of mechanical fatigue, calcification, glycation, lipid peroxidation and digestion with proteases. The fragmented peptides (EDP) are bioactive and retain the VGVAPG peptide sequence found in tropoelastin, which gives them a similar signaling capacity. Blood levels of EDP are elevated in people with obliterating atherosclerosis of the legs, type IIb hyperlipidemia, hypertension, diabetes mellitus, and ischemic heart disease. It is not known if the elevated EDP levels are the result or the cause of the disease. Recent studies in mice suggest that although fragmentation of elastic fibers may be secondary to a genetic or acquired disease, circulating EDP contributes to the further progression of the disease (Duque et al., 2018).

Sequestration of TGF-β

TGF-β is secreted from cells in the form of a large latent complex (LLC) that contains a latency related peptide (LAP) and a latent TGF-β binding protein (LTBP). Latent TGF-β binding protein locates latent TGF-β in the extracellular matrix. LTBP1 and LTBP3 bind well to all three isoforms of TGF-β (TGF-β1, TGF-β2, TGF-β3). TGF-β1 signaling is altered when arterial elastic fibers are not properly folded and by proteolytic cleavage of proteins associated with elastic fibers, indicating a bidirectional effect between TGF-β1 activity and elastic fiber integrity (Cocciolone et al., 2018).

Elastin-Based Composite Fiber Scaffolds

Elastin can be synthesized in combination with other proteins or polymers to enhance the structural and functional properties of a number of biomaterials. As with pure elastin materials, the so-called electrospinning of composite structures based on elastin leads to the formation of a three-dimensional structure with a large specific surface area and high porosity. This technique allows the production of scaffolds that mimic the structure and architecture of the extracellular matrix. Electrospinning is widely used in the manufacture of tubular and flat scaffolds for applications in the skin, bones, nerves, and especially vascular tissues.

Elastin and Photoaging

Elastosis is a form of degenerative disease in which there is an accumulation of elastin in the tissues. There are many causes, but the most common is actinic elastosis of the skin, also known as solar elastosis, which is caused by prolonged and excessive exposure to the sun, a process known as photoaging (Weihermann et al., 2017).

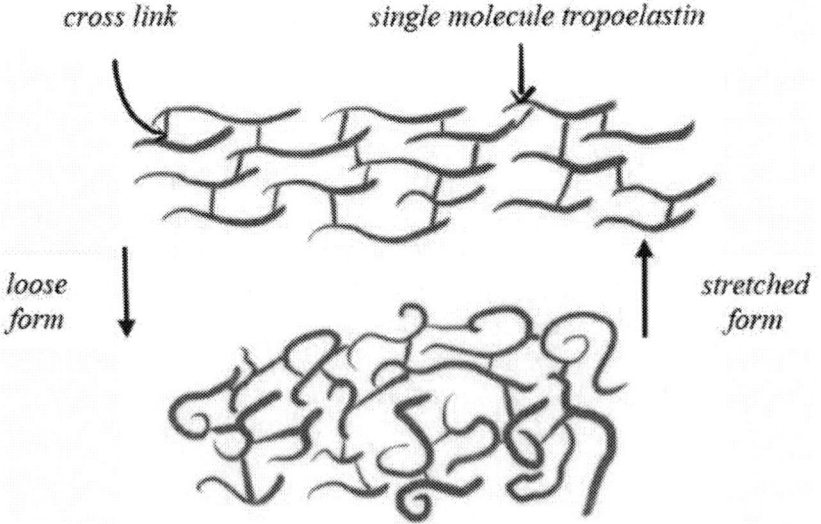

Figure 1. Two forms of elastin - loose and stretched (Own elaboration).

Summarizing the work, it can be concluded that elastin and elastic fibers play a key mechanical role on many levels. Elastin is a unique protein due to its limited synthesis time, long half-life, multi-stage assembly process and a high degree of cross-linking, which provides mechanical flexibility to large arteries.Improperly assembled or degraded elastic fibers change the passive mechanical properties of the artery wall. The consequences extend far beyond physical changes in elastic fibers to the initiation of several mechanisms of arterial remodeling, including modulation of the VSMC phenotype, extracellular matrix deposition, inflammatory cell infiltration and EDP release that stimulate cytokine signaling and activate inflammatory cells. Inflammatory cells, in turn, secrete proteases which contribute to additional degradation of elastic fibers in the positive feedback cycle.

Despite increasing therapeutic opportunities to control diseases such as hypertension, obesity and diabetes, the elastic fiber damage that often accompanies these diseases is now irreparable, meaning that cardiovascular risks from arterial stiffness may persist even when other conditions are present. Aspects of the disease are controlled. The paper also presents numerous applications of elastin, including as a signal molecule or an element of a composite scaffold, which enables the increase of structural and functional properties of a number of biomaterials. Elastin has also been shown to be associated with photoaging and the occurrence of SVAS congenital heart disease – supravalvular aortic stenosis.

References

Cocciolone AJ, Hawes JZ, Staiculescu MC, Johnson EO, Murshed M, Wagenseil JE. Elastin, arterial mechanics, and cardiovascular disease. *Am J Physiol Heart Circ Physiol*, 2018 Aug 1; 315(2):H189-H205.

Duque Lasio ML, Kozel BA. Elastin-driven genetic diseases. *Matrix Biol,* 2018 Oct; 71-72:144-160.

Vindin H, Mithieux SM, Weiss AS. Elastin architecture. *Matrix Biol,* 2019 Nov 84: 4-16.

Wang Y, Song EC, Resnick MB. Elastin in the Tumor Microenvironment. *Adv Exp Med Biol,* 2020; 1272:1-16.

Weihermann AC, Lorencini M, Brohem CA, de Carvalho CM. Elastin structure and its involvement in skin photoageing. *Int J Cosmet Sci,* 2017 Jun; 39(3):241-247.

Chapter 4

Collagen

Julia Buszek, Dorota Bartusik-Aebisher[*] and David Aebisher

Medical College of The University of Rzeszów, Rzeszów, Poland

Abstract

There are almost 30 types of collagen in our body. Preclinical studies have shown that hydrolyzed collagen stimulates the regeneration of collagen tissue by increasing not only collagen synthesis, but also the synthesis of smaller components (glycosaminoglycans and hyaluronic acid). Clinical studies show that continuous intake of hydrolyzed collagen helps reduce and prevent joint pain, bone loss and skin aging. This study shows that the correct amount and quality of collagen present in our body have a significant impact on the comfort of our lives.

Keywords: collagen, osteoarthritis, hyaluronic acid, fibrous protein

To date, more than 20 different types of collagen have been identified that together form a family. Their main function is to create networks and structures of the extracellular matrix, as well as basement membranes. The occurrence and function vary depending on the type of collagen. Type I collagen is the most common type of collagen in the human body. It dominates mature tendons and ligaments and is the main structural protein in the human body (Bolke et al., 2019). Its fibrils are formed by packing triple peptide helices. It creates a variety of tissue scaffolds by attaching to other molecules

---

[*] Corresponding Author's Email: dbartusikaebisher@ur.edu.pl.

In: The Biochemical Guide to Proteins
Editors: David Aebisher and Dorota Bartusik-Aebisher
ISBN: 979-8-88697-493-5
© 2023 Nova Science Publishers, Inc.

in various proportions. Type I collagen is a triple helix with a length of 300 nm. 3 parallel polypeptide chains wound around each other join to form fibrils (Figure 1). Type I alpha-2 collagen (SMM) improves type I collagen synthesis, cell proliferation, cell migration and elastin synthesis, supporting a significant anti-wrinkle effect. Type V collagen, also known as COLV, forms fibers and is a regulatory collagen. It has a minimum of three other molecular isoforms – α1 (V) 2 α2 (V), α1 (V) 3 and α1 (V) α2 (V) α3 (V) – formed by combinations of three different α-α1 (V) polypeptide chains, α2 (V) and α3 (V). COLV is a relatively small collagen of the extracellular matrix. COLV interacts with matrix collagens and structural proteins to impart structural integrity to tissue scaffolds. Binds matrix macromolecules, modulating behavior and cellular functions (Bolke et al., 2019).

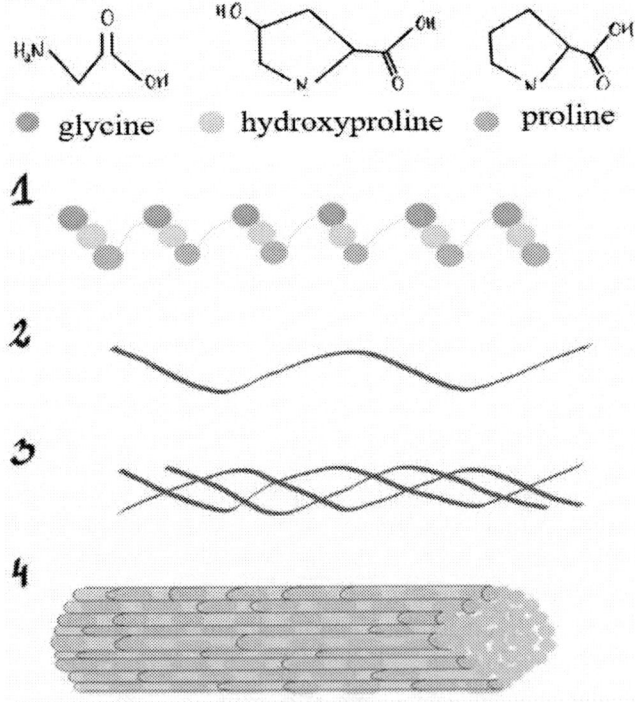

Figure 1. Collagen structure 1. Amino acid sequence 2. Single chain amino acid sequence 3. Triple helix 4. Collagen fiber [Own elaboration; source of inspiration for drawing: Review of the Applications of Biomedical Compositions Containing Hydroxyapatite and Collagen Modified by Bioactive Components - Scientific Figure on ResearchGate. Available from: https://www.researchgate.net/figure/Schematic-representation-of-collagens structure_fig1_351059436 accessed 20 Dec, 2021].

Collagens mediate important hemostasis, influencing the integrity and stability of the vessel wall. Varying collagen turnover related to its uncontrolled formation or degradation may result in pathological conditions, for example with fibrosis. The thickening of the vessel wall due to the excess amount of collagen can lead to obstruction of the arteries or thrombosis. Conversely, thinner vessel walls resulting from collagen degradation or deficiency can lead to rupture of the vessel wall or an aneurysm. Prevention of excessive haemorrhage or thrombosis relies on collagen-mediated actions. Von Willebrand factor, integrins and glycoprotein VI, as well as clotting factors, can bind collagen to restore normal haemostasis after injury (Daneault et al., 2017).

Human type I alpha-2 collagen stimulates collagen synthesis, wound healing and elastin production in normal human skin fibroblasts (HDF). Skin aging overlaps with internal processes – genetic and hormonal factors, and external – external environment, including constant exposure to light, chemicals and toxins. These factors reduce the synthesis of type I collagen and elastin in fibroblasts, as well as increase melanin in melanocytes. Type I collagen is the most abundant type of collagen and the main structural protein in the tissues of the human body. Collagen in tandem with elastin are necessary to maintain the skin structure in the dermis, and additionally to give it elasticity. Moreover, collagen, a fibrous protein, plays a major role in strengthening the mechanical strength of the skin and makes up most of the dermis. Type I collagen mutations are associated with aortic dilatation, dissection, and rupture, arterial dilatation and rupture, and osteogenesis imperfect (Karna et al., 2020).

Type III collagen is the second most abundant type of collagen contained in the human body. Additionally, it is a key structural component of the organization of fibrous collagen. Type III collagen is often combined with type I collagen to form heterotypic type I/III fibrils. Although frequent combinations of type I and III collagens have been demonstrated, clustering of type III and II collagen has also been observed. The frequent occurrence of type III collagen is associated with many functions it performs, and any deficiencies and abnormalities have various symptoms. The existence of an autosomal dominant human COL3A1 mutation is associated with impaired production of type III collagen (reduced amounts or defective type III collagen). Patients with this mutation suffer from Ehlers-Danlos syndrome (VEDS, vEDS, EDS type IV). Due to the amount of this type of collagen in the human body, a mutation in its gene is associated with many symptoms, e.g., greater susceptibility to fatal rupture of vessels and organs, symptoms of premature aging, degeneration of the musculoskeletal system. In addition, it has been shown that the deficiency of the same collagen in mice causes symptoms such as deterioration of the quality of skin wound healing during aging, affects the development of the neocortex, as well as the development and repair of the skeleton (Meyer et al., 2019).

The V-type collagen mentioned in the introduction, otherwise known as COLV, gives structural integrity to tissue scaffolds and binds matrix macromolecules, modulating cell behavior and functions. Additionally, COLV binds to COLI (type I collagen) into heterotypic fibrils in the cornea and dermis, acting as the dominant regulator of collagen fibrillogenesis. Type V collagen deficiency is associated with the loss of corneal transparency and the classic Ehlers-Danlos syndrome. In contrast, COLV overexpression occurs in cancer, granulation tissue, inflammation, atherosclerosis, and fibrosis of the lungs, skin, kidneys, adipose tissue and liver. The COLV isoform, which has an α3 (V) chain, is involved in mediating the functions of islet cells. In the liver, COLV is a small but repeating component of the extracellular matrix. An excess of COLV is associated with early as well as advanced liver fibrosis. COLV neoepitopes have been shown to be a useful non-invasive serum biomarker for assessing the progression and regression of fibrosis in experimental liver fibrosis (Rittié et al., 2017).

Collagen is a compound that is abundant in the body, there are almost 30 types of collagen. Multiple functions testify to its great importance in the human body, but also not only (it is also used in graveyard, cosmetics, and cooking). All disorders related to the synthesis, operation, mutations and connecting with other compounds to form complexes can carry many abnormalities and diseases. In a study of the beneficial effects of taking hydrolyzed collagen, it has been shown to reduce the damage and consequences of collagen loss, such as joint pain and erosion (osteoarthritis), loss of bone density (osteoporosis), and skin aging. Preclinical studies have shown that hydrolyzed collagen stimulates the regeneration of collagen tissue by increasing not only collagen synthesis, but also the synthesis of smaller components (glycosaminoglycans and hyaluronic acid). Clinical studies show that continuous intake of hydrolyzed collagen helps reduce and prevent joint pain, bone loss and skin aging. These results, as well as the high level of tolerance and safety, make the intake of hydrolyzed collagen an incentive for long-term use in degenerative diseases of bones and joints and in the fight against skin aging. This study shows that the correct amount and quality of collagen present in our body have a significant impact on the comfort of our lives. Taking it may have positive effects in the form of better well-being, prevention of diseases related to its deficiency and skin aging.

References

Bolke L., Schlippe G., Gerß J., Voss W. A Collagen Supplement Improves Skin Hydration, Elasticity, Roughness, and Density: Results of a Randomized, Placebo-Controlled, Blind Study. *Nutrients.* 2019 Oct 17;11(10):2494.

Daneault A., Prawitt J., Fabien Soulé V., Coxam V., Wittrant Y. Biological effect of hydrolyzed collagen on bone metabolism. *Crit Rev Food Sci Nutr.* 2017 Jun 13;57(9):1922-1937.

Karna E., Szoka L., Huynh T. Y. L., Palka J. A. Proline-dependent regulation of collagen metabolism. *Cell Mol Life Sci.* 2020 May;77(10):1911-1918.

Meyer M. Processing of collagen based biomaterials and the resulting materials properties. *Biomed Eng Online.* 2019 Mar 18;18(1):24.

Rittié L. Method for Picrosirius Red-Polarization Detection of Collagen Fibers in Tissue Sections. *Methods Mol Biol.* 2017;1627:395-407.

Chapter 5

Myosin

Dominik Jaklik, Dorota Bartusik-Aebisher[*] and David Aebisher
Medical College of The University of Rzeszów, Rzeszów, Poland

Abstract

Myosin is a protein widespread in living organisms it occurs in lower organisms: amoebas, nematodes, fungi, but also in plants and animals. So far, over a dozen classes of myosin have been discovered that are involved in processes such as: the conversion of electrical signals into a nerve impulse, exocytic and endocytic transport, cell division, transport of cytoplasm between cells, contraction of muscle cells. Type II myosin is the one that is responsible for the possibility of muscle cell contraction, thanks to the ability to hydrolyze ATP and bind to actin.

Keywords: myosins, phototransduction, endocytic transport, cell division, muscle cell, amoebas, nematodes, fungi

Introducton

Myosins are present not only in the muscle cells of humans and animals, they are also present in plants, nematodes, amoebas and fungi. They can be found both in the cytoplasm of cells and in their nuclei. They constitute a group of motor proteins responsible for the mobility of processes taking place in cells,

---

[*] Corresponding Author's Email: dbartusikaebisher@ur.edu.pl.

In: The Biochemical Guide to Proteins
Editors: David Aebisher and Dorota Bartusik-Aebisher
ISBN: 979-8-88697-493-5
© 2023 Nova Science Publishers, Inc.

and are most often associated with their role in the contraction of skeletal and smooth muscle (Schröder et al., 2020). However, there are several classes of myosin, not all of which have been described to date. It is a "family" of proteins that are actin-dependent, capable of binding and hydrolyzing ATP (thanks to the presence of ATPase), thereby gaining energy which allows them to slide along the thin filament from the minus end to the plus end (Titus et al., 2018). Moreover, they have a common domain structure scheme. As in the most popular of myosins (II), and in others, we can distinguish a heavy chain with a motor domain at the amino terminus (actin and ATP binding), then we have a neck with 1-7 places where light chains attach, followed by a tail (carboxyl terminus), which is the most diverse zone, which determines the specificity of a given type of myosin. Some of the myosins in the heavy chain tails have sequences of different lengths, which makes it possible to dimerize them (presence of 2 heavy chains) (Bugyi et al., 2020). Types of myosin:

- Myosin I: ubiquitous, absent in plants, has a single head that interacts with actin filaments, differs in the structure of the tail, which determines the cell component transported by a given class, acts as a monomer in vesicular transport
- Myosin II: is a dimer, has two heads, the tails form myosin filaments, present in the muscle cells of most animal cells where it is involved in muscle contraction
- Myosin III, believed to be involved in phototransduction, i.e., the conversion of light into electrical signals in the suppositories and rods of the retina
- Myosin V, composed of 2 heavy and 6 light chains, is found in the brain, melanocytes, testes, kidneys, lungs, liver, heart as well as in epithelium and glandular tissue, and is involved in the transport of cargo from the center of the cell to its border.
- Myosin VI: is a dimer or monomer involved in exo- and endocytosis, cell adhesion and migration, e.g., in the hair cells of the inner ear
- Myosin VII: exists as 2 isoforms, myosin VIIA has a longer heavy chain, is found in the middle ear, retina, testes and kidneys, while myosin VIIB is present in the epithelium of the kidneys and intestines
- Myosin VIII, present in plant cells, is involved in cell division as well as in the flow of cytoplasm between cells
- Myosin IX: present in tissue organisms, beginning with the nematodes, as the 2 isoforms IXA and IXB

- Myosin X: found in filopodia in mammals
- (cytoplasmic projections)
- Myosin XI, present in plant cells (plastids and mitochondria), takes part in the movement of chloroplasts, creates connections between them. Takes part in proper hair growth.
- Myosin XIV: the plasma membrane of intracellular parasites involved in the invasion of cells
- Myosin XV: contributes to the development of the actin core of cilia in the inner ear
- Myosin XVIII: present in humans in the form of 2 isofromes, XVIIIA in striated muscles, hematopoietic cells of the bone marrow, small intestine, XVIIIB in the thymus, pancreas, prostate
- Class IV, XII, XIII, XVI, XVII myosins are still present, but their role has not yet been thoroughly investigated.

As mentioned earlier, myosin II forms the so-called thick filaments. They are capable of shifting actin filaments in relation to each other (Sweeney et al., 2020). Myosin II has two heads that bind to actin. Each has its own orientation and moves the thin filaments in a different direction, i.e., the heads act in opposite directions, but always move towards the positive end of the actin filament (Squire et al., 2019). During muscle contraction, oppositely directed actin filaments overlap each other. Generally, muscle cell contraction consists in the cyclic attachment and detachment of myosin heads from actin threads, the transverse bridges are broken. As a result of combining these proteins, conformational changes in the head take place (due to the separation of ADP), which results in the formation of a force impulse shifting actin threads against myosin (Guhathakurta et al., 2018). The role of myosin in the muscle systolic-diastolic cycle in stages:

- In the diastolic phase: myosin heads hydrolyze ATP-> ADP + Pi, myosin is formed in the high-energy conformation
- When the muscle is stimulated to contract, the actin filaments are exposed, which allows the head to bind and form the actin-myosin-ADP-Pi complex
- The resulting complex causes the separation of Pi and the formation of a force impulse, as a result of which the actin thread is shifted through the head of the myosin, which in turn becomes low-energy

- Subsequently, ATP attaches to the myosin head, which causes a lower affinity for actin, which allows the fibers to relax.

Myosin is a protein widespread in living organisms, as explained above, it occurs in lower organisms: amoebas, nematodes, fungi, but also in plants and animals. They are proteins present in the form of monomers or dimers, which depends on the number of heavy chains in the molecule. So far, over a dozen classes of myosin have been discovered that are involved in processes such as: the conversion of electrical signals into a nerve impulse (eye retinal cells), exocytic and endocytic transport, cell division, transport of cytoplasm between cells, contraction of muscle cells, or even in invasion of the body by parasitic cells. In humans, it is most often associated with the occurrence in muscle cells, in which it forms thick filaments, however, it is worth knowing that in addition to myocytes of muscle tissue, it is present in: hematopoietic cells of the bone marrow, small intestine, prostate, thymus, pancreas, testes, lungs, kidneys, melanocytes in the brain, retina, inner ear hair cells. Type II myosin is the one that is responsible for the possibility of muscle cell contraction, thanks to the ability to hydrolyze ATP and bind to actin, resulting in the formation of the actin-myosin complex, followed by actin being pulled up by the myosin head (contraction) and subsequently they disconnect and the muscle cell relaxes.

References

Bugyi B, Kengyel A. Myosin XVI. *Adv Exp Med Biol,* 2020; 1239:405-419.
Guhathakurta P, Prochniewicz E, Thomas DD. Actin-Myosin Interaction: Structure, Function and Drug Discovery. *Int J Mol Sci,* 2018; 19(9):2628.
Schröder RR. The Structure of Acto-Myosin. *Adv Exp Med Biol,* 2020; 1239:41-59.
Squire J. Special Issue: The Actin-Myosin Interaction in Muscle: Background and Overview. *Int J Mol Sci,* 2019; 20(22):5715.
Sweeney HL, Houdusse A, Robert-Paganin J. Myosin Structures. *Adv Exp Med Biol,* 2020; 1239:7-19.
Titus MA. Myosin-Driven Intracellular Transport. *Cold Spring Harb Perspect Biol,* 2018; 10(3):a021972.

Chapter 6

Fibronectin

Jacek Mazanek, Dorota Bartusik-Aebisher[*] and David Aebisher
Medical College of The University of Rzeszów, Rzeszów, Poland

Abstract

Since the 1940s, thanks to scientists, the protein fibronectin has been known. Getting to know its exact structure allowed to understand how the reactions related to fibronectin proceed. Due to its structure, fibronectin can exist in a soluble and insoluble form depending on the situation and place of occurrence. The globular form of fibronectin is found in blood plasma, amniotic fluid, cerebrospinal fluid, and saliva, and is soluble, while the fibrillar form is found in the extracellular matrix or associated with the cell membrane and is insoluble and therefore supports. A deficiency or lack of fibronectin due to genetic diseases in utero can cause birth defects as well as impaired or completely abolished ability to heal wounds.

Keywords: fibronectin, glycosaminoglycans, laminin, Schwann cells, blood plasma

Fibronectin belongs to glycoproteins and is one of the components of the extracellular matrix. It belongs to the non-collagen matrix proteins, along with elastin, laminin, thrombospondin, tenascin, matriline, nidogen, fibulin, and fibrillin. It is produced and secreted by fibroblasts, monocytes, endothelial

---

[*] Corresponding Author's Email: dbartusikaebisher@ur.edu.pl.

In: The Biochemical Guide to Proteins
Editors: David Aebisher and Dorota Bartusik-Aebisher
ISBN: 979-8-88697-493-5
© 2023 Nova Science Publishers, Inc.

cells and various types of epithelial cells along with keratinocytes, as well as by astrocytes, Schwann cells, chondrocytes and peritoneal macrophages. Hepatocytes are mainly responsible for the synthesis and secretion of plasma fibronectin. This protein is present in the walls of blood vessels, in the basal membranes, between smooth muscle cells, sarcolemia of striated muscles, walls of the liver sinusoids and loose fibrous connective tissue (Lin et al., 2019. Fibronectin first appears during embryogenesis on the sinus node of the blastula. It then degrades or is redistributed along with the mesenchymal cells that differentiate into muscle, cartilage, and tubular epithelium. This protein is involved in organogenesis, organism development, cell adhesion and migration, maintaining homeostasis, angiogenesis, vascular repair and remodeling. It mediates the attachment of fibrin or collagen to macrophages. Acting in tandem with integrin receptors, fibronectin initiates a cascade of events that will lead to the transmission of signals from the external environment to the interior of the cell, regulating the organization of the cytoskeleton and cell function. The absence or deficiency of fibronectin in the cellular matrix can cause birth defects and a reduced or completely abolished ability to heal wounds (Lin et al., 2019).

In 1948, Peter Morrison, John Edsall, and Susan Miller discovered a water-insoluble protein in fractionating human plasma and called it cold-insoluble globulin. In the 1970s, Hynes and Vahery discovered a protein on the surface of cells that Vahery called fibronectin in order to standardize the nomenclature. Fibronectin is a dimer consisting of two similar polypeptide chains with similar molecular weights of 235-250 kDa. Each of the fibronectin chains is composed of repeating amino acid motifs, arranged irregularly. There are 12 kinds of type I modules, 2 kinds of type II modules and 15-17 kinds of type III modules. Type I and II are made up of approximately 40 amino acids each. Type I amino acids form two planes of the antiparallel β structure, one with two and the other with three strands, while in type II two orthogonal, double-stranded planes of the antiparallel β structure. Both contain four cysteine residues, which allows the formation of disulfide bridges, which stiffen the protein structure. Type III is composed of approximately 90 amino acids and consists of two opposing planes of the antiparallel β structure, three-stranded and four-stranded. Type III does not form disulfide bonds. There are three sites of alternative splicing in fibronectin. The use or omission of an exon leads to the inclusion or exclusion of one of the two type III repeats – EDB (EIIIB) located between the fibronectin III7 and III8 repeats, and EDA (EIIIA) located between the fibronectin III11 and III12 repeats. The third area of alternative splicing is located in the non-homologous stretch called the V region between repeats III14 and III15 and contains the α4β1 integrin binding site (Patten et al., 2021).

In fibronectin, repeating modules build domains, i.e., sites of interaction with other molecules, such as collagen, heparin, fibrin, extracellular matrix proteins, other cells, and glycosaminoglycans. These modules enable fibronectin to perform its functions. Fibronectin is soluble in plasma, amniotic fluid, cerebrospinal fluid, saliva, and in non-soluble forms it is bound to a cell membrane or to a macromolecular cross-linked form in the extracellular matrix. In solutions, fibronectin is a globular protein, and when incorporated into the matrix, the fibronectin molecule stretches into a fibrillar form (Qiao et al., 2020).

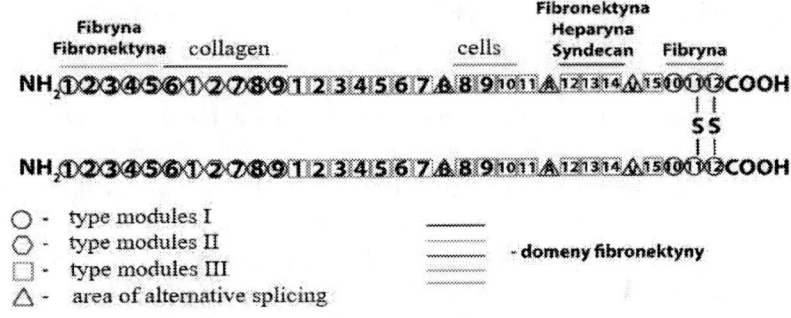

Source: Own study based on Fibronectin repair function.

Figure 1. Structure of fibronectin.

Fibronectin mediates many repair processes: skin, bones, heart valves, corneal wound healing and the proliferation of peripheral neurites. Wound healing occurs in three stages: local inflammation, proliferation and remodeling. Fibronectin works at all stages of wound healing. Plasma fibronectin is involved in the initial stage of wound healing, helping to form a clot. It has a built-in V region in only one dimer subunit, which allows it to be bound to a fibrin clot. The first step in binding fibronectin to fibrin is a non-covalent bond between the fibrin I domain (I1-I5) and a specific site in the fibrin molecule. Under the influence of factor XIIIa (fibrin stabilizing factor), a covalent bond is formed that stabilizes and strengthens the complex. Domain I (I10-I12) then reacts with fibrin. This interaction is weaker, but allows it to assume the appropriate conformation of fibronectin. This facilitates adhesion, the spread of fibroblasts and the migration of cellular elements into the wound (Son et al., 2017).

Fibronectin contains domains that bind to the appropriate type of receptor, which is integrins. The integrins thus mediate the contact between the cell and the extracellular matrix, in the contact between cells, and some participate in both types of information transfer. The RGD sequence of fibronectin binds to the integrins α3β1, α5β1, α8β1, αIIbβ3, and the LDV sequence mainly binds to the integrin α4β1. They are the best known fibronectin binding integrins (Speziale et al., 2019).

Since the 1940s, thanks to scientists, the protein fibronectin has been known. Thanks to their discovery, we are aware of the important functions it plays in the human body. Getting to know its exact structure allowed to understand how the reactions related to fibronectin proceed. Dimer structure and interactions between the chains are related to the amino acid composition of the protein. Domains, i.e., repeating modules in the fibronectin chain, are the places of interaction of a protein molecule with other molecules, e.g., other cells, heparin, glycosaminoglycans, matrix proteins or fibrin. Due to its structure, fibronectin can exist in a soluble and insoluble form depending on the situation and place of occurrence. The globular form of fibronectin is found in blood plasma, amniotic fluid, cerebrospinal fluid, and saliva, and is soluble, while the fibrillar form is found in the extracellular matrix or associated with the cell membrane and is insoluble and therefore supports. Such a structure of the protein allows fibronectin to perform a repair function. It allows binding to the fibrin clot, which facilitates adhesion, the spread of fibroblasts and the migration of cellular elements into the wound. In contrast, the interaction of fibronectin with integrins is necessary for the transmission of information from the extracellular matrix to the cell or between cells. It regulates the organization of the cytoskeleton and the functioning of the cell. Therefore, a deficiency or lack of fibronectin due to genetic diseases in utero can cause birth defects as well as impaired or completely abolished ability to heal wounds.

References

Lin T C, Yang C H, Cheng L H, Chang W T, Lin Y R, Cheng H C. Fibronectin in Cancer: Friend or Foe. *Cells.* 2019 Dec 20;9(1):27.

Patten J, Wang K. Fibronectin in development and wound healing. *Adv. Drug Deliv. Rev.* 2021 Mar;170:353-368.

Qiao P, Lu ZR. Fibronectin in the Tumor Microenvironment. *Adv. Exp. Med. Biol.* 2020;1245:85-96.

Son M, Miller E.S. Predicting preterm birth: Cervical length and fetal fibronectin. *Semin. Perinatol.* 2017 Dec;41(8):445-451.

Speziale P, Arciola C R, Pietrocola G. Fibronectin and Its Role in Human Infective Diseases. *Cells.* 2019 Nov 26;8(12):1516.

Chapter 7

Colony-Stimulating Factors (CSFs)

Adrian Groele, Dorota Bartusik-Aebisher[*] and David Aebisher

Medical College of The University of Rzeszów, Rzeszów, Poland

Abstract

Colony stimulating factors are extremely important substances in the human body that contribute to the formation and proliferation of white blood cells. Their number should be equal to the number of their physiological occurrence, because both too low and too high their concentration carries serious consequences. Many see them as an opportunity to treat cancer, and due to the fact that they naturally increase the leukocyte population, they are a worth considering method of fighting this serious disease. It will be the most natural and minimally invasive method that will be able to provide patients with greater comfort of therapy in the future.

Keywords: colony stimulating factors (CSF), multiple colony stimulating factor (MULTI-CSF), granulocyte-macrophage colony stimulating factor GM-CSF), aacute myeloid leukemia (AML)

Colony stimulating factors, or CSF for short, are glycoproteins. This means that they are made of protein and saccharides connected with each other by a glycosidic bond. The name "colony stimulating factors" comes from the method how they were discovered. Blood stem cells were grown on a semi-

---

[*] Corresponding Author's Email: dbartusikaebisher@ur.edu.pl.

In: The Biochemical Guide to Proteins
Editors: David Aebisher and Dorota Bartusik-Aebisher
ISBN: 979-8-88697-493-5
© 2023 Nova Science Publishers, Inc.

solid matrix, this culture was designed to prevent the movement of single, proliferating cells. It was necessary to obtain separate, immiscible colonies in the study. The newly formed communities were first treated with various substances, and then it was examined which types of colonies were stimulated by the factors given to them.

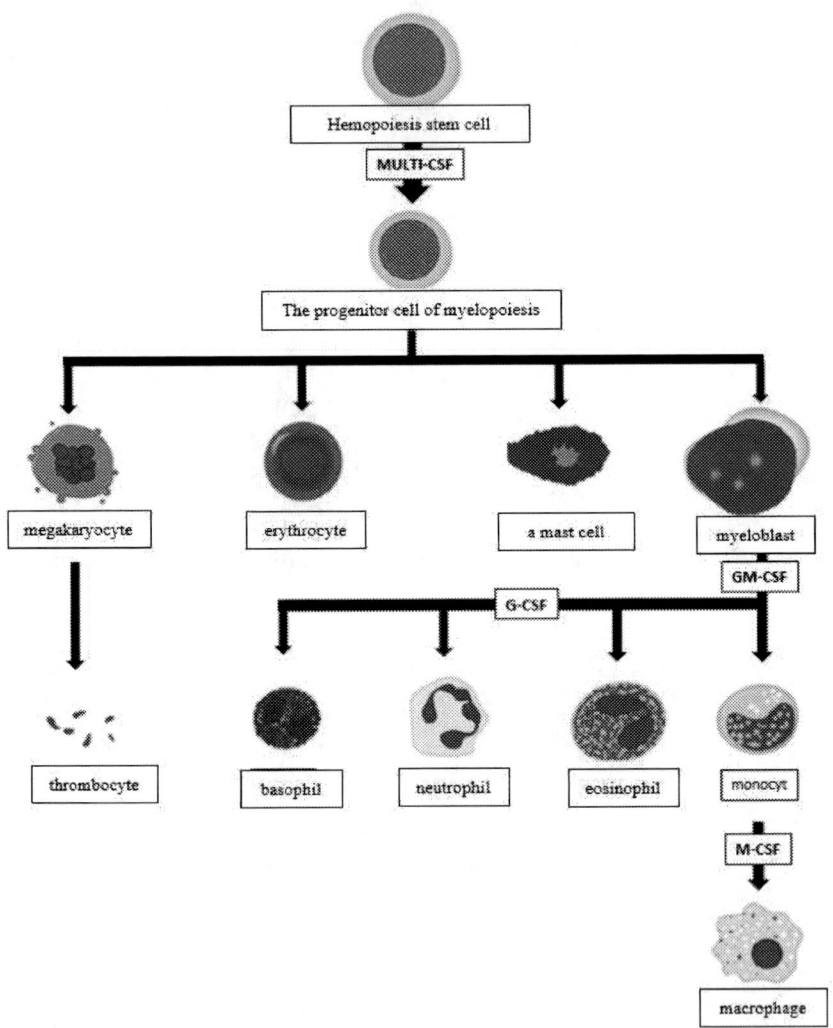

Figure 1. Myelopoiesis, taking into account the activity of factors stimulating the growth of appropriate cells [Own elaboration].

The names of each type of CSF were defined during the course of the trial (Singhal et al. 2019). A group that stimulates the formation of granulocytes colony (G-CSF), which is involved in the production of polynuclear leukocytes (neutrophils, eosinophils, basophils), has been distinguished. Another group is the granulocyte-macrophage colony stimulating factor (GM-CSF), which stimulates the development of granulocytes and monocytes. Then there is the macrophage colony stimulating factor (M-CSF), which induces the differentiation of macrophage precursors. Multiple colony stimulating factor (MULTI-CSF)/interleukin 3 (IL-3) induces bone marrow stem cells to differentiate into bone marrow progenitor cells. The mechanism and type of cells affected by the factors are given in the diagram below. Each type of stimulating factor binds to receptor proteins (CSFR) that are found on the surface of hematopoietic cells (Singhal et al. 2019). In this way, intracellular signaling pathways are activated, and it is this process that causes the proliferation and differentiation of cells, usually into white blood cells. Leukocytes are involved in human immune responses, so their proper growth is crucial for the proper functioning of the body (Jeannin et al. 2018).

CSF, i.e., proteins covalently linked with oligosaccharides, are glycoproteins with a mass of 18-70 kDa. This factor is active in both glycosylated and non-glycosylated forms (Jeannin et al. 2018). The half-life of glycosylated CSFs is longer than that of non-glycosylated CSFs, however the difference is not much, usually around 1 to 6 hours. Stimulating factors are the products of most tissues and cells of the immune system (Tan et al. 2018). During their induction by other cells or endotoxins, CSF can increase their production up to 1000 times. Thus, in the presence of an inducing agent, CSF is produced quickly and at a high level and can therefore function as a sensitive system controlling and regulating the formation of hematopoietic cells. The exception is M-CSF, which is produced in higher concentrations but with retained stability (Barros Pinto et al. 2020). In 1967 it was shown that CSF is necessary for the survival of progenitor cells as well as their progeny. In the years 1977-1987, when the human marrow was further analyzed and purified, CSF became more accessible, which made it possible to estimate that in addition to supporting the development of granulocyte and macrophage colonies, the development of erythroid cells, eosinophils, megakaryocytes, mast cells and T and B lymphocytes could also be supported. In contrast, in 1990, it was demonstrated that the absence of CSF would result in death from apoptosis. CSF also has the ability to stimulate the activity of mature cells (Barros Pinto et al. 2020). For example, GM-CSF has the ability to act on mature neutrophils. It affects their chemotaxis, stimulates metabolism, enhances the response to microorganisms and causes increased production of regulatory proteins in neutrophil cells. Similar effects have been documented for G-CSF, M-CSF and IL-3 which act on monocytes and macrophages. On the membranes of blood cells, receptors for CSF are present, which show specific regions - signaling chains (Barros Pinto et al. 2020). They have the capacity to initiate the various events necessary to elicit diverse biological responses. They transmit signals important for the maturation and proliferation of myeloid progenitor cells. Too much CSF can have a bad effect on the body. On the example of the mice tested, it was shown that overexpression of GM-CSF means an increase in the number of granulocytes and macrophages, it is associated with various lethal inflammatory lesions in the muscles, intestines and lungs. Also, excessive levels of IL-3 lead to the formation of too many mast cells and hematopoietic cells, resulting in excessive scratching and itching (Barros Pinto et al. 2020). This could be due to degranulation of skin mast cells. Overproduction of G-CSF does not lead to any particular changes, however, with inactivation of the cytokine signaling gene 3 (Socs3) suppressor, even a simple amount of G-CSF can cause paralysis and even death through increased accumulation of neutrophils in the spinal cord, bone marrow, lungs or liver. Based on research in the early 1970s, it was established that chronic leukemia (CML) cells, which formed apparently normal granulocyte colonies, and non-proliferating acute myeloid leukemia (AML) cells, remained dependent on stimulation of proliferation by the material in which it was located. CSF. On this basis, conclusions were drawn that CSF may be a cofactor in the development of myeloid leukemia. Neoplastic cells receive proliferative stimuli from CSF. However, a more prominent role for CSF factors, especially G-CSF and GM-CSF, is in increasing the low level of white blood cells in cancer patients after chemotherapy. The above factors trigger the release of hematopoietic stem cells into the blood. Thanks to this process, they have replaced the bone marrow as populations more effective for transplanting patients whose treatment has resulted in bone marrow damage.

Colony stimulating factors are extremely important substances in the human body that contribute to the formation and proliferation of white blood cells. Their number should be equal to the number of their physiological occurrence, because both too low and too high their concentration carries serious consequences. Deficiency results in a reduced number of mature cells such as granulocytes and macrophages, and their insufficient amount may cause various diseases due to the lack of immune cells. On the other hand, an increased amount of CSF may have a carcinogenic effect, resulting from the excessive work of stimulating factors. It is also important that the number of each type of CSF is appropriate to the current needs of cells in the body. That is, G-CSF for neutrophils, eosinophils and basophils, M-CSF for macrophages, and also GM-CSF for granulocytes and MULTI-CSF for marrow progenitor cells (Tan et al. 2018). Numerous studies conducted by many scientists over the last decades show and prove the importance of these factors. Many see them as an opportunity to treat cancer, and due to the fact that they naturally increase the leukocyte population, they are a worth considering method of fighting this serious disease. It will be the most natural and minimally invasive method that will be able to provide patients with greater comfort of therapy in the future. Despite many experiments with CSF, there are still many undiscovered possibilities for which CSF proteins can be used.

References

Barros Pinto MP, Marques G. The effects of granulocyte colony-stimulating factors (G-CSFs) in leucocytes. *Hematol. Oncol. Stem Cell Ther*. 2020; 13(1): 40-41.

Huang X, Hu P, Zhang J. Genomic analysis of the prognostic value of colony-stimulating factors (CSFs) and colony-stimulating factor receptors (CSFRs) across 24 solid cancer types. *Ann. Transl. Med*. 2020;8(16): 994.

Jeannin P, Paolini L, Adam C, Delneste Y. The roles of CSFs on the functional polarization of tumor-associated macrophages. *FEBS J*. 2018; 285(4): 680-699.

Singhal A, Subramanian M. Colony stimulating factors (CSFs): Complex roles in atherosclerosis. *Cytokine*. 2019; 122: 154190.

Tan Y, Shuai C, Wang T. Critical Success Factors (CSFs) for the Adaptive Reuse of Industrial Buildings in Hong Kong. *Int. J. Environ. Res. Public Health*. 2018; 15(7): 1546.

Chapter 8

Histones

Agnieszka Zaleszczyk, Dorota Bartusik-Aebisher* and David Aebisher

Medical College of The University of Rzeszów, Rzeszów, Poland

Abstract

Histones are proteins with many functions, but the most important are those that build DNA. Histones are proteins with a small molecular weight: 11.3-21.0 kDa. They are rich in basic amino acids, mainly arginine and lysine. They are also water-soluble proteins. There are also many tissue-specific variants of histones, especially in the case of linker histone, for example CenH3 – responsible for the organization of centromeres and kinetochores, H3.3 increases transcription activity, H2A.X – facilitates repair and recombination of DNA. Histones constitute 25-40% of the chromatin mass.

Keywords: histones, drug-induced lupus, proteins, DNA, lysine

The way in which genetic information is stored varies from organism to organism, but one thing is certain – all information is stored on DNA. 3.079 billion base pairs is quite a lot, and we put them in the body in a twisted, 2 meter long spiral of DNA. And so in each cell, which in total gives us several million kilometers. We set off for the sun, we come back, and the DNA is even longer than our journey (Zhou et al. 2019).

---

* Corresponding Author's Emal: dbartusikaebisher@ur.edu.pl.

In: The Biochemical Guide to Proteins
Editors: David Aebisher and Dorota Bartusik-Aebisher
ISBN: 979-8-88697-493-5
© 2023 Nova Science Publishers, Inc.

Each cell has to pack DNA in such a way that there is still space for other organelles and the size of the cell itself is small. In eukaryotes, most of the DNA is located in the nucleus, while in prokaryotes, in the cytoplasm. In the early 1970s, the genetic material was examined under a microscope, and to the surprise of the researchers, they saw "breads on a string." In the following years, it was proved that the string is DNA and the beads are histone proteins. Histones are a group of proteins commonly found in the nuclei of eukaryotic cells, characterized by significant evolutionary stability. The most important equation for this separation is: DNA + histone proteins = nucleosome. Now, let me just mention that the histone part consists of 8 histone molecules, and that DNA consists of 146 base pairs (Zhou et al. 2019). There are five types of histones:

- H1 – the most basic and the largest of the histones, known as the linker histone,
- H2A,
- H2B,
- H3,
- H4, these are core histones.

Histones are proteins with a small molecular weight: 11.3-21.0 kDa. They are rich in basic amino acids, mainly arginine and lysine. They are also water-soluble proteins (Yang et al. 2021).

These proteins are modular and consist of three parts: COOH-terminal, central and NH2-terminal. The core of the nucleosome is made up of some COOH and the central part, while some of the NH2 remains outside. Histones are among the slowest changing proteins. The second subfamily is the linker histones, present in more variants and evolving faster than the other histones. These proteins constitute a buckle that binds the whole of Histones, regardless of their species source, they are divided into 5 types: H1 (very rich in lysine), H2A and H2B (rich in lysine) as well as H3 and H4 (rich in arginine). The most conserved are histones H3 and H4, which differ slightly even in evolutionarily distant species (Yang et al. 2021).

For example, the H4 histone of peas and calf thymus differ in only 2 amino acids.

The primary function of histone H1 and H5 is to stabilize the compact structure of the nucleosome. The lack of linker histones changes the structure of chromatin, which becomes loose, and the nucleosomes that build it lack the characteristic structure. Histone H1 differs significantly from the others. It is much larger, more alkaline and shows a large species and even tissue diversity (Yang et al. 2021). H1 histones commonly found in all types of nucleated cells and H5 histones specific for avian and amphibian nucleated erythrocytes.

These proteins, like core histones, are made up of three domains. The central part (also known as the globular part) is made of about 80 amino acids and is characterized by the highest conservativeness. The globular domain consists of a bundle of three α-helices and a β-harmonica (in this case called a wing) located near the C-terminus. Due to this structure, linker histones can be included in the HTH protein family, although typical HTH proteins such as CAP (catabolic activator protein) have a four-amino acid twist between the second and third helix, which is absent in histones. To DNA, and differences in the amino acid composition between histones H1 and H5 result in a different affinity of these forms for DNA (Doolin et al. 2020).

A nucleosome consists of a protein "cluster" wrapped externally by a DNA strand. Each histone cluster is cylindrical with a diameter of 10 nm and a height of 6 nm, containing 2 molecules of each of the histones: H2A, H2B, H3 and H4 (octamer). Supplemented with an extra piece of the coil (2 full turns, 166 bp in total) and H1 becomes the chromatosome, Truncating the double chain of the chromatosome below 166 bp deprives it of its ability to bind H1 (Nadal et al. 2018).

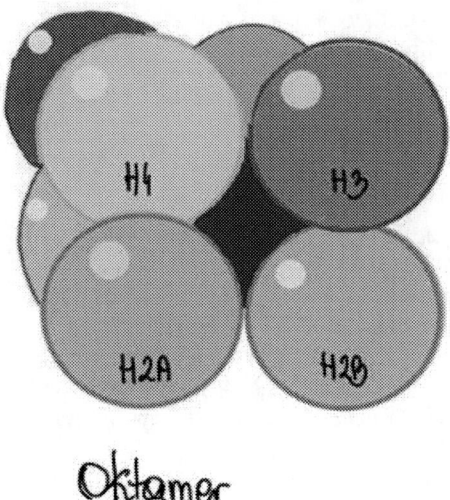

Figure 1. Graphical representation of the octamer - visible 4 pairs of histones: H2A, H2B, H3 and H4 [Own study].

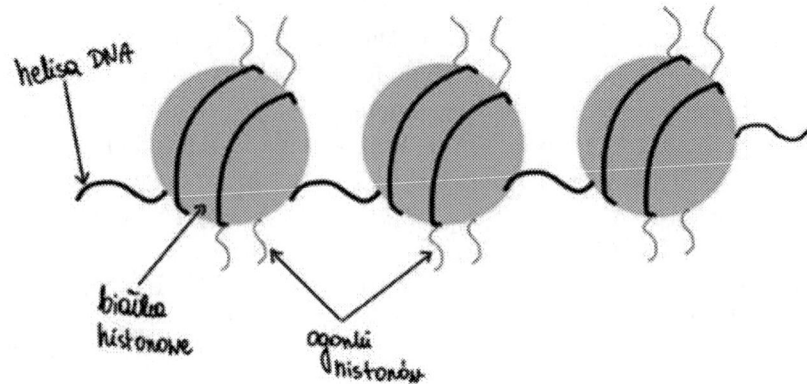

Figure 2. Graphical representation of histones with a wound DNA chain [Own study].

In summary, histones are proteins with many functions, but the most important are those that build DNA. Histones H3 and H4 are the most evolutionarily conserved, while histone H1 is the most variable. There are also many tissue-specific variants of histones, especially in the case of linker histone, for example CenH3 – responsible for the organization of centromeres and kinetochores, H3.3 increases transcription activity, H2A.X – facilitates repair and recombination of DNA. Histones constitute 25-40% of the chromatin mass. Drug-induced lupus is a disease associated with the formation of anti-histone autoantibodies (Zhang et al. 2021).

References

Doolin T, Gross S, Siryaporn A. Physical Mechanisms of Bacterial Killing by Histones. *Adv. Exp. Med. Biol.* 2020;1267:117-133.

Nadal S, Raj R, Mohammed S, Davis BG. Synthetic post-translational modification of histones. *Curr. Opin. Chem. Biol.* 2018 Aug;45:35-47.

Yang G, Yuan Y, Yuan H, Wang J, Yun H, Geng Y, Zhao M, Li L, Weng Y, Liu Z, Feng J, Bu Y, Liu L, Wang B, Zhang X. Histone acetyltransferase 1 is a succinyltransferase for histones and non-histones and promotes tumorigenesis. *EMBO Rep.* 2021 Feb 3;22(2):e50967.

Zhang Y, Sun Z, Jia J, Du T, Zhang N, Tang Y, Fang Y, Fang D. Overview of Histone Modification. *Adv. Exp. Med. Biol.* 2021;1283:1-16.

Zhou BR, Bai Y. Chromatin structures condensed by linker histones. *Essays Biochem.* 2019 Apr 23;63(1):75-87.

Chapter 9

Protamines

Agnieszka Gancarz, Dorota Bartusik-Aebisher[*] and David Aebisher

Medical College of The University of Rzeszów, Rzeszów, Poland

Abstract

Protamines are basic proteins because they contain large amounts of arginine, so they are cationic and can bind, for example, with sulphates. Their secondary structure may take the form of α-helices, β-turns and other structures not stabilized by intermolecular hydrogen bonds. It has been shown that abnormalities in genes containing protamine information can lead to male infertility due to the fact that chromatin in male reproductive cells does not reach the proper consistency. Protamine used as antagonism to heparin, so its administration prevents the anticoagulant effect of this substance, which makes it useful in preventing bleeding during cardiac, vascular and other operations – low molecular weight protamines are used.

Keywords: protamines, sulphates, anaphylactoid reactions, salmin, DNA, X-ray

Inroduction

Protamines are small basic proteins that contain high amounts of arginine (60-80%). They are used clinically to prevent hemorrhages, after cardiovascular

---

[*] Corresponding Author's Email: dbartusikaebisher@ur.edu.pl.

In: The Biochemical Guide to Proteins
Editors: David Aebisher and Dorota Bartusik-Aebisher
ISBN: 979-8-88697-493-5
© 2023 Nova Science Publishers, Inc.

operations, because they are heparin antagonists (it has an anticoagulant effect) – protamines are cationic, so by binding to sulphate groups they counteract the anticoagulant activity of heparin (patients are given it in the form of sulphate protamine). However, this use can be dangerous as it can trigger a non-immune response mediated by the complement system to release thromboxane and histamine (anaphylactoid reactions) and via an immunoglobulin-mediated pathway (anaphylactic reactions). The former are less harmful, while the latter significantly threaten the health and life of the patient (Steger et al., 2018).

Protamines are also involved in the condensation of chromatin in sperm along with nuclear and transitional proteins. In the first stage of this process, somatic histones are replaced with transitional proteins (TP-1, TP-2), which are then converted into protamines (P-1, P-2) in the process of spermatids elongation, the resulting chromatin is highly condensed and does not undergo transcription. Genes containing information about P-1, P-2 are mapped on the 16p13.3 chromosomes, they contain only one intron, moreover, they share the TP-2 locus. Abnormal expression of protamine-encoding genes can cause male infertility and spermiogenesis errors by defective chromatin condensation in the male reproductive cell (Steger et al., 2018).

Figure 1. Sperm.

The first protamine was isolated from salmon semen. This was done by degreasing the sperm and then extracting it with dilute hydrochloric acid. It was formed as a precipitate – the double salt of platinum, was not bound to sulfur or phosphorus, was not soluble in water, but dissolved when combined with excess hydrochloric acid. The isolated protamine is called salmin, its content in dry sperm mass is 26.8%. A similar protein was obtained from carp seed. It was initially used to make NPH insulin, and then began to be used as a heparin antagonist in the prevention of postoperative bleeding (Bahl et al., 2017).

Protamines have been shown to contain very few types of amino acids, as most are known to be basic amino acids such as arginine (60-80% of all amino acids), which is the basic building block of protamines, and there are also histidine or lysine (maybe all three at once). In addition to arginine, all proteins also contain alanine and serine. The structure may also include proline, valine (mostly), glycine, isoleucine (many) and threonine, aspartic and glutamic acids (Bahl et al., 2017).

In the human HP2 protamine, it was found that a linear array of five positively charged arginine residues is formed on one side of the peptide, which most likely facilitates the protamine function in interacting with negatively charged phosphate groups during DNA packaging in the cell (Kleene et al., 2018). In salmin and protamine obtained from squid seed associated with DNA, the secondary structure is in the form of an α-helix, β-turns and other structures unstabilized by intermolecular hydrogen bonds are also present, while the presence of β-harmonica is not recorded. In salmin – about 20% of α-helices, 40% of other structures, in squid protamine – 40% of α-helices and 20% of other structures, in both β-turns are approximately 40% (Kleene et al., 2018). The conformation of the protamine resembles that of the elongated left-handed helix of polyproline II. She was examined by X-ray and CD analysis. It is sensitive to changes in pH and temperature, but it is stable when it comes to the action of ions, detergents and alcohols. Protamines used as heparin antagonists are proteins with low molecular weight – average weight 1100 Daltons.

Protamine is used as a heparin antagonist because it counteracts its anticoagulant properties, it is a component of NPH insulin, and its function is also participation in the condensation of chromatin in the emerging sperm. NPH insulin is an insoluble human insulin suspension, classified as an intermediate-acting insulin. It is used in type 1, 2 and gestational diabetes. It is a basic insulin and is widely used. Name: N - neutral, P - protamine, H - creator's name (Hagedorn). It is obtained by precipitation of human (recombinant) insulin with zinc and with the participation of protamine at a neutral pH (protamine to insulin ratio - 1: 5), resulting in the formation of an insulin-protamine complex. Due to the binding of insulin with protamine, it can diffuse into the subcutaneous tissue, where, due to the action of cleaving enzymes and macrophages, insulin is gradually released from the complex, which causes its prolonged action in the body and delayed absorption (Sarkar et al., 2017).

Protamine is used to neutralize the anticoagulant properties of heparin in cardiovascular surgery. Upon its administration, unfractionated heparin forms an aggregate of the heparin-protamine salt and does not fulfill its function. Protamine is administered intravenously, but due to the possibility of an immune system reaction (there may be a sensitivity to this protein, especially in people who used NPH insulin), first a test with a small amount of it and watch for signs of allergy should be performed (Sarkar et al., 2017). You should also exercise caution in setting the dose, as too high a dose may be counterproductive. Protamine has a relatively fast duration of action – it forms a compound with heparin within 5 minutes and its half-life is approximately 10 minutes. Salt metabolism is not well understood, it may be metabolised by the liver or the kidney (Hao et al., 2019).

Protamines are basic proteins because they contain large amounts of arginine, so they are cationic and can bind, for example, with sulphates. Besides arginine, all protamines also contain alanine and serine. Their secondary structure may take the form of α-helices, β-turns and other structures not stabilized by intermolecular hydrogen bonds. They naturally participate in the condensation of chromatin in male sperm along with nuclear and transitional proteins – histones are converted into transitional proteins, and these are replaced with protamines. It has been shown that abnormalities in genes containing protamine information can lead to male infertility due to the fact that chromatin in male reproductive cells does not reach the proper consistency. Thanks to their properties, they are also used in clinical medicine. Due to the fact that they can form a complex with insulin, they are used to form NPH insulin – the basal insulin used in both types of diabetes (type I, type II) and in gestational diabetes, which has a prolonged action in the body and delayed absorption (this insulin is classified as for intermediate-acting insulins). Another application of protamines is their antagonism to heparin, so its administration prevents the anticoagulant effect of this substance, which makes it useful in preventing bleeding during cardiac, vascular and other operations – low molecular weight protamines are used.

References

Bahl K, Senn JJ, Yuzhakov O, Bulychev A, Brito LA, Hassett KJ, Laska ME, Smith M, Almarsson Ö, Thompson J, Ribeiro AM, Watson M, Zaks T, Ciaramella G. Preclinical and Clinical Demonstration of Immunogenicity by mRNA Vaccines against H10N8 and H7N9 Influenza Viruses. *Mol Ther*, 2017; 25(6):1316-1327.

Hao SL, Ni FD, Yang WX. The dynamics and regulation of chromatin remodeling during spermiogenesis. *Gene*, 2019; 706:201-210.

Kleene KC. Gordon Dixon, protamines, and the atypical patterns of gene expression in spermatogenic cells. *Syst Biol Reprod Med*, 2018; 64(6):417-423.

Sarkar M, Prabhu V. Basics of cardiopulmonary bypass. *Indian J Anaesth*, 2017; 61(9):760-767.

Steger K, Balhorn R. Sperm nuclear protamines: A checkpoint to control sperm chromatin quality. *Anat Histol Embryol*, 2018 47(4):273-279.

Chapter 10

Enkephalins as Peptides Widely Distributed in the Body

Natalia Guzik, Dorota Bartusik-Aebisher[*] and David Aebisher
Medical College of The University of Rzeszów, Rzeszów, Poland

Abstract

Enkephalins are a group of compounds that are widely distributed in the body and fulfill their functions by stimulating opioid receptors in various parts of the body. Met-enkephalin is also known as growth factor (OGF) because it is involved in the regulation of the immune system. Met-enkephalin is involved in tissue regeneration and activates the OGF receptor. The conducted research also shows the presence of met-enkephalin and its precursor in the sperm head, where it acts as an endogenous mediator of sperm motility. Enkephalins are therefore compounds that have a wide range of applications, therefore they are often used as drugs in the course of many diseases.

Keywords: met-enkephalin, leu-enkephalin, amino acid, cerebrospinal fluid (CSF)

Opioids are a group of compounds that show affinity to opioid receptors found in various brain structures, in the spinal cord and on cells of peripheral tissues (including the gastrointestinal tract, urinary system, vas deferens). In 1973,

---

[*] Corresponding Author's Email: dbartusikaebisher@ur.edu.pl.

In: The Biochemical Guide to Proteins
Editors: David Aebisher and Dorota Bartusik-Aebisher
ISBN: 979-8-88697-493-5
© 2023 Nova Science Publishers, Inc.

Simon et al. Terenius and Pert and Snyder obtained biochemical evidence for the existence of stereospecific opiate binding sites, or opiate receptors, in the brain of animals. This discovery led them to conclude that endogenous opioids must be produced in the human brain. In 1985, Kosterlitz detected the presence of enkephalins in preparations from the brain tissue of pigs and later in the cerebrospinal fluid of humans are peptides classified as endogenous opioids, characterized by high biological activity. Enkephalins have an analgesic effect (Bando et al. 2017). They fulfill their function while affecting the circulatory, immune and nervous systems. The source of enkephalins is the pituitary and adrenal medulla. Enkephalins are found mainly in the striatum, hypothalamus and midbrain. Two of them – met-enkephalin and leu-enkephalin – belong to the group of peptide neurotransmitters and neuromediators. Met-enkephalin is an agonist at the zeta and delta opioid receptor and to a lesser extent the mu opioid receptor. Met-enkephalin has analgesic, neuromodulatory, immunomodulating, anti-inflammatory, antinociceptive and gastrointestinal motility modulating properties. Leu-enkephalin is an agonist at the delta-opioid and mi-opioid receptor. Leu-enkephalin, in addition to its role in neurotransmission, is also an analgesic and a human and rat metabolite (Bando et al. 2017).

Three researchers: Terenius, Wahlstrom and Kosterlitz together with their colleagues in the years 1974-1975 detected substances in human cerebrospinal fluid (CSF) with activity similar to exogenous opioids. There were 2 pentapeptides: Met-enkephalin and Leu-enkephalin, the amino acid sequence of which was virtually identical, except for the last amino acid (Dou et al. 2020).

Both enkephalins are produced from the precursor proenkephalin A (PENK), encoded by the PPE gene (prepro-enkephalin gene). The PPE gene determines the expression of a protein consisting of 267 amino acid residues. PENK is a conserved protein that is sensitive to CP1 and CP2 convertases. It is worth mentioning that research shows the effect of morphine on a significant change of transcription in the PPE gene. It seems interesting that short-term, repeated administration of nicotine increases the synthesis and processing of proenkephalin As a result of the proenkephalin A cleavage reaction carried out by proteases, 4 Met-enkephalin molecules and 1 Leu-enkephalin molecule, a heptapeptide composed of Met-enkephalin and two amino acid residues (arginine and phenylalanine residues), an octapeptide in which to form Met-enkephalin are formed. Enkephalins are attached with 3 amino acid residues (arginine, glycine, and leucine residues) (Kainov et al. 2020).

Table 1. Compiled by Natalia Guzik

	Met-enkephalin	**Leu-enkephalin**
Summary formula	$C_{27}H_{35}N_5O_7S$	$C_{28}H_{37}N_5O_7$
The last amino acid	Methionine	Leucine
Amino acid sequence	Tyr-Gly-Gly-Phe-Met	Tyr-Gly-Gly-Phe-Leu

Enkephalins are widely distributed in the human body and are found wherever opioid receptors are present in the cells. They can be found in the Auerbach plexus (gastrointestinal muscle), thalamus, in the gray matter of the peri-duct, posterior horns of the spinal cord, striatum, hypothalamus, midbrain, in the limbic system, cortex, pituitary gland, as well as in cells such as fibroblasts and keranocytes.

Enkephalins have a strong stimulating effect on opioid receptors, among which there are three types: MOR (mu, μ), DOR (delta, δ) and KOR (kappa, κ) receptors. All classic opioid receptors share a common origin. They belong to the family of monobotropic receptors (associated with G protein), which are composed of 7 transmembrane domains, 6 loops, glycosylated N-terminus outside and C-terminus present inside the cell. Extracellular fragments show great variability in terms of amino acid sequence and spatial conformation, which is related to selective ligand binding. Activation of these receptors results in the opening of potassium channels, an extracellular current of potassium ions, hyperpolarization of the cell membrane and closure of calcium channels, which results in the inhibition of synaptic transmission. Met-enkephalin and leu-enkephalin have the highest affinity for DOR receptors, and lower for MOR and KOR receptors (Lykke-Andersen et al. 2021).

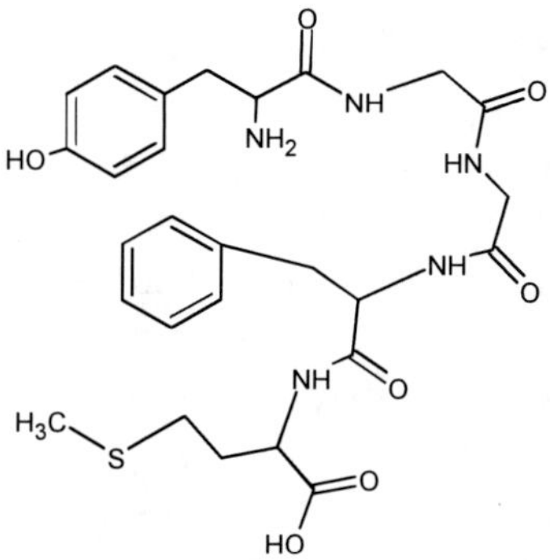

Figure 1. Structure of Met-enkephalin.

The activity of Met- and Leu-enkephalin is widely represented in the human body. They are included in the group of neurotransmitters and neuromodulators. Their main task is to reduce the level of pain, which is caused by their binding to the δ-type opioid receptors in the CNS. Their analgesic effect is compared to that of morphine. Additionally, the participation of these endogenous opioids in the stress response has been demonstrated – they have an anxiolytic effect promoted by the endocrine system. Due to the presence of DOR receptors in the gastrointestinal tract, enkephalins affect the motility of the digestive system, pancreas and carbohydrate metabolism. They can also lower blood pressure and body temperature. Changes in the opioid system are associated with the development of hypertension. Enkephalins are responsible for suppressing the heart's activity. Met-enkephalin levels increase in the heart muscle as chronic heart failure worsens. Met-enkephalin is called opioid growth factor (OGF) and has an inhibitory effect on the growth of pancreatic cancer, hepatoblastoma, breast cancer, ovarian cancer, and lung cancer. The activity of enephalins also causes behavioral changes in humans, such as: motor agitation, appetite, and convulsions (Yin et al. 2020).

Enkephalins are a group of compounds that are widely distributed in the body and fulfill their functions by stimulating opioid receptors in various parts of the body. They are found not only in the central nervous system, where they act as neuromediators and neuromodulators released from the post- and presynaptic terminals of neurons. In the digestive system, enkephalins are responsible for inhibiting intestinal peristalsis, secretion of gastric and pancreatic juices, bile and sphincter contraction. It is worth mentioning that reduced secretion of enkephalins by the gastrointestinal tract and leukocytes occurs in people with disturbed intestinal hormone release and in patients with inflammation of the intestine. Reducing the level of enkephalins in the limit is associated with the development of hypertension, because these opioid peptides in the circulatory system are involved in lowering blood pressure and internal temperature. Met-enkephalin is also known as growth factor (OGF) because it is involved in the regulation of the immune system. In the organs of the immune system, it is responsible for stimulating the growth and maturation of immune cells, influences the migration of monocytes to peripheral tissues and their conversion to macrophages, lymphocytes, and neutrophils. It is also important that Met-enkephalin is involved in tissue regeneration and activates the OGF receptor. The conducted research also shows the presence of met-enkephalin and its precursor in the sperm head, where it acts as an endogenous mediator of sperm motility. Enkephalins are therefore compounds that have a wide range of applications, therefore they are often used as drugs in the course of many diseases.

References

Bando, S., Nishikado, A., Hiura, N., Ikeda, S., Kakutani, A., Yamamoto, K., Kaname, N., Fukatani, M., Takagi, Y., Yukiiri, K., Fukuda, Y., and Nakaya, Y. (2018). Efficacy and safety of rivaroxaban in extreme elderly patients with atrial fibrillation: Analysis of the Shikoku Rivaroxaban Registry Trial (SRRT). *J Cardiol.*, Feb, 71(2), 197-201.

Dou, Y., Kalmykova, S., Pashkova, M., Oghbaie, M., Jiang, H., Molloy, K. R., Chait, B. T., Rout, M. P., Fenyö, D., Jensen, T. H., Altukhov, I., and LaCava, J. (2020). Affinity proteomic dissection of the human nuclear cap-binding complex interactome. *Nucleic Acids Res.*, Oct 9, 48(18), 10456-10469.

Kainov, Y. A., and Makeyev, E. V. (2020). A transcriptome-wide antitermination mechanism sustaining identity of embryonic stem cells. *Nat Commun.*, Jan 17, 11(1), 361.

Lykke-Andersen, S., Rouvière, J. O., and Jensen, T. H. (2021). ARS2/SRRT: at the nexus of RNA polymerase II transcription, transcript maturation and quality control. *Biochem Soc Trans.*, Jun 30, 49(3), 1325-1336.

Yin, J., Kim, S. S., Choi, E., Oh, Y. T., Lin, W., Kim, T. H., Sa, J. K., Hong, J. H., Park, S. H., Kwon, H. J., Jin, X., You, Y., Kim, J. H., Kim, H., Son, J., Lee, J., Nam, D. H., Choi, K. S., Shi, B., Gwak, H. S., Yoo, H., Iavarone, A., Kim, J. H., and Park, J. B. (2020). ARS2/MAGL signaling in glioblastoma stem cells promotes self-renewal and M2-like polarization of tumor-associated macrophages. *Nat Commun.*, Jun 12, 11(1), 2978.

Chapter 11

Platelet-Derived Growth Factor (PDGF)

Monika Błądek, Dorota Bartusik-Aebisher[*] and David Aebisher
Medical College of The University of Rzeszów, Rzeszów, Poland

Abstract

> Platelet growth factor, a glycoprotein present in several isoforms, is undoubtedly an important factor found in the human body, which is produced by a variety of cells and is widely expressed. Since its discovery the several decades ago, many different studies have been carried out, which allowed to learn about its structure, biosynthesis process and operation. The emergence of different varieties of this factor confirms that it performs many important and different functions. Platelet growth factor is still a factor that is discussed in subsequent scientific research, which is a source of new, valuable and sometimes surprising information. Undoubtedly, the knowledge about this factor will keep increasing, as it is a topic worth further attention.

> **Keywords:** platelet-derived growth factor (PDGF), oligodendrocyte progenitor cells (OPCs), fibrosarcoma, microblastic skin tumors

PDGF, or platelet growth factor, is one of the growth factors that regulate cell growth and division. It was discovered in the 1980s as a serum growth factor for fibroblasts, smooth muscle cells and glial cells. It is synthesized by many different cell types and is widely expressed. The synthesis of this factor occurs in response to external stimuli such as low oxygen pressure or stimulation by

---

[*] Corresponding Author's Email: dbartusikaebisher@ur.edu.pl.

In: The Biochemical Guide to Proteins
Editors: David Aebisher and Dorota Bartusik-Aebisher
ISBN: 979-8-88697-493-5
© 2023 Nova Science Publishers, Inc.

other cytokines and growth factors. It plays a significant role in the formation of blood vessels, the growth of pre-existing blood vessels, mitogenesis and chemotaxis. PDGF is a dimer composed of two polypeptide units linked together by a disulfide bridge. It may consist of various combinations of subunits, e.g., PDFG-AA, PDFG-AB, PDGF-BB. PDGF plays a key role during development. Studies on the effects of PDGF on animal development have shown that PDGFR-α plays an important role in gastrulation and development of the nerve crest, skull, heart, central nervous system, skeleton, gonads, lungs, intestines, and skin. There is also evidence that it performs physiological functions during adult life. Increased platelet-derived growth factor activity has been linked to certain diseases and pathologies, namely certain gliomas, sarcomas and leukemias. PDGF also drives pathological responses in vascular disorders such as atherosclerosis, pulmonary hypertension, and retinal disease. Establishing the causal action of PDGF in some diseases may enable their treatment by the use of PDGF antagonists (Cadamuro et al., 2019).

PDGF is a highly basic (pI-10.2) glycoprotein that has been isolated in 2 equally active protein fractions differing only in the content of covalently bonded carbohydrates. PDGF I weighs 31kD and contains approximately 7% bound carbohydrate. The mass of PDGF II is 28kD and contains 4% carbohydrates. The mitogenic activity, amino acid composition, and immunological reactivity of PDGF I and PDGF II are substantially equal. Both forms appear to have 18 half cystine residues, all located in a disulfide bond. The reduction of these disulfide bonds causes the biological activity to be abolished and the PDGF to be split into larger and smaller chains. There are 5 different isoforms of this factor that activate the cellular response through two different types of receptors. Among the known ligands, we distinguish PDGF-AA (PDGFA), PDGF-BB (PDGFB), PDGF-CC (PDGFC), PDFG-DD (PDGF) AND PDGF-AB, which is a heterodimer of PDGFA and PDGFB (Medamana et al., 2017). Encoded by the PDGFA gene. The expression product of this gene may exist as a PDGFB-related homodimer or heterodimer. Two possible splicing variants were identified for the gene encoding this subunit.

Subunit encoded by the PDGFB gene. For the gene encoding this subunit, also two possible splicing variants can be distinguished. Mutations due to reciprocal translocations between chromosomes 22 and 17 where the PDGFB and COL1A1 genes are located, or alternatively by the presence of an abnormal small ring chromosome connecting the two genes and giving rise to a fusion gene, result in anomalies in the production of this subunit (Figure 1).

The formation of the fusion gene results in overproduction of the molecule, which is believed to be the reason for the development and progression of protein-related fibroblastic and microblastic skin tumors: giant cell fibroma and nodular skin fibrosarcoma (Lee et al., 2018).

This protein is encoded by the PDGFC gene. It consists of 345 amino acids. The expression product of this gene only forms homomers. Unlike PDGFA and PDGFB, it has an N-terminal CUB domain. PDGFC is involved in the development of the palate and morphogenesis of the integument. This protein is encoded by the PDGFD gene. The expression product of this gene can only form homodimers, which makes it impossible to dimerize it with other subunits of this family. 2 possible splicing variants have also been identified for this gene. Płytkowy czynnik wzrostu syntezowany jest przez różne rodzaje komórek. Synteza tego czynnika następuje w odpowiedzi na stymulacje zewnętrzną taką jak niskie ciśnienie tlenu, trombiny bądź poprzez stymulację przez inne cytokininy lub czynniki wzrostu (Roskoski et al., 2018).

Platelet growth factor is mitogenic in the early stages of development, leading to proliferation of undifferentiated mesenchymal cells and some populations of progenitor cells. In the later stages of development, it plays a role in tissue remodeling and cell differentiation, and in inductive events related to pattern formation and morphogenesis. PDGF is an important mitogenic factor for connective tissue and especially for fibroblasts involved in the wound healing process. The important role this factor plays in wound healing is due to the fact that PDGF is a factor that allows the cell to skip checkpoints in the G1 phase of the cell cycle and consequently divide. Another role this factor plays is to maintain the proliferation of oligodendrocyte progenitor cells (OPCs). It has been shown that fibroblast growth factor (FGF) is also involved in this process, which activates the pathway of positive regulation of PDGF receptors in oligodendrocyte progenitor cells (Shen et al., 2020).

Figure 1. A simplified diagram of the structure of the PDGF-BB monomer (no side groups are shown in the diagram).

Platelet growth factor, a glycoprotein present in several isoforms, is undoubtedly an important factor found in the human body, which is produced by a variety of cells and is widely expressed. Since its discovery several decades ago, many different studies have been carried out, which allowed to learn about its structure, biosynthesis process and operation, which provided scientists with valuable information allowing for a better understanding of the functioning of the complex mechanism of the human body. The acquired knowledge also allowed us to learn the etiology of some diseases that are associated with mutations that lead to abnormalities in the formation of different varieties of this factor, which allows for the development of better methods of treating these diseases. The above short description of platelet growth factor shows that it plays an important role both during fetal development and later in life of an individual, of which its most important role seems to be participation in mitogenic and proliferative processes. The emergence of different varieties of this factor confirms that it performs many important and different functions. Platelet growth factor is still a factor that is discussed in subsequent scientific research, which is a source of new, valuable and sometimes surprising information. This fact is confirmed by the number of entries on the factor that have been shown and still appear since 1978 in scientific databases. Undoubtedly, the knowledge about this factor will keep increasing, as it is a topic worth further attention.

References

Cadamuro M., Brivio S., Mertens J., Vismara M., Moncsek A., Milani C., Fingas C., Cristina Malerba M., Nardo G., Dall'Olmo L., Milani E., Mariotti V., Stecca T., Massani M., Spirli C., Fiorotto R., Indraccolo S., Strazzabosco M., Fabris L. Platelet-derived growth factor-D enables liver myofibroblasts to promote tumor lymphangiogenesis in cholangiocarcinoma. *J Hepatol.* 2019 Apr;70(4):700-709.

Lee C., Li X. Platelet-derived growth factor-C and -D in the cardiovascular system and diseases. *Mol Aspects Med.* 2018 Aug;62:12-21.

Medamana J., Clark R. A., Butler J. Platelet-Derived Growth Factor in Heart Failure. *Handb Exp Pharmacol.* 2017;243:355-369.

Roskoski R. Jr. The role of small molecule platelet-derived growth factor receptor (PDGFR) inhibitors in the treatment of neoplastic disorders. *Pharmacol Res.* 2018 Mar;129:65-83.

Shen S., Wang F., Fernandez A., Hu W. Role of platelet-derived growth factor in type II diabetes mellitus and its complications. *Diab Vasc Dis Res.* 2020 Jul-Aug;17(7):1479164120942119.

Chapter 12

Immunoglobulins

Maria Dycha, Dorota Bartusik-Aebisher[*] and David Aebisher
Medical College of the University of Rzeszów, Rzeszów, Poland

Abstract

Immunoglobulins are glycoproteins secreted by plasma cells in response to the presence of an antigen. There are 5 types of immunoglobulins in the human body: IgM, IgG, IgA, IgE, and IgD. Each type is made up of four polypeptide chains: two light and two heavy, which are linked by a disulfide bridge. IgG is the highest percentage of plasma antibodies and is synthesized in the secondary immune response. IgA is found mainly in secretions, e.g., in saliva, tears and genital secretions, and IgE plays a role in the body's allergic response and protection against parasites. Immunoglobulins play an essential role in the body's defense against pathogens, and thanks to their specificity and flexibility, they are an attractive option for the development of therapy against in the different diseases.

Keywords: immunoglobulin, Kawasaki disease, systemic lupus erythematosus (SLE), organ transplants, combined immune disorders (SCID)

In 1890, two scientists, Emil von Behring and Kitasato Shibasaburō, informed about the existence of a factor in the blood capable of neutralizing diphtheria

---

[*] Corresponding Author's Email: dbartusikaebisher@ur.edu.pl.

In: The Biochemical Guide to Proteins
Editors: David Aebisher and Dorota Bartusik-Aebisher
ISBN: 979-8-88697-493-5
© 2023 Nova Science Publishers, Inc.

toxin. After over 100 years of research, these factors have become one of the best known structurally, functionally and genetically groups of proteins – immunoglobulins, with a wide range of properties. Immunoglobulins (Ig), otherwise known as antibodies, are glycoproteins that make up 20% of plasma proteins. They are secreted by plasma cells in the course of the humoral immune response that appears as a result of the presence of antigens from other organisms, e.g., bacteria and viruses. Plasma cells are cells of the immune response, formed as a result of stimulation of B lymphocytes by specific immunogens (bacteria, proteins). Immunogens, reacting with the BCR receptor on the surface of B lymphocytes, generate a signal directing the activation of transcription of factors capable of stimulating the synthesis of antibodies that are highly specific for the B cell stimulating immunogen. Moreover, one B cell clone forms a specific immunoglobulin (Bracken et al. 2019).

Circulating antibodies have the ability to recognize an antigen in tissue fluids and serum, and the immune system to remember it, thanks to developing memory cells. It is a type of B lymphocytes characterized by a long life and the ability to quickly produce large amounts of antibodies in a short time, which contributes to the initiation of a specific mechanism of immune memory. In 1939, Tiselius and Kabat used electrophoresis to separate the immunized serum into albumin, alpha-goblin, beta-globulin and gamma-globulin fractions. It allowed to distinguish 5 types of immunoglobulins in the human body: IgM, IgG, IgA, IgE, IgD (Basyal et al. 2021).

Apart from functional similarity, i.e., the ability to react in a specific way with antigens, all types of antibodies are characterized by a similar structure. They are peptide – two light (L) and two heavy (H), bound by disulfide bonds and arranged in a light-heavy-heavy-protein Y-shaped arrangement, composed of four light chains (Figure 1).

The separation of different types of immunoglobulins results, inter alia, from the differentiation of the first domains (with a free amino group) of the heavy or light chain, as they are regions with a variable sequence of amino acids. The H and L domains are responsible for the complementarity of the antigen binding site with the antigenic determinant and thus determine the specificity of the antibody's reaction with specific antigens. There are two functional parts of antibodies: the Fab fragment, which contains the antigen-binding site, and the Fc fragment, which mediates biological functions. Each chain is arranged in a so-called immunoglobulin tangles, called domains, heavy in four or five and light in two (Lu et al. 2020).

IgM is a pentamer with a molecular weight of 970 kD, and its mean serum concentration is 1.5 mg/ml. This type of immunoglobulin is produced mainly in the primary immune response to infectious agents or antigens, and can also activate the classical complement pathway. IgM is considered a strong agglutinin and IgM monomer is used as the B cell receptor (BCR).

IgG is a monomer with a molecular weight of 150 kD and a serum concentration of 9.0 mg/ml. This type of immunoglobulin is bivalent, meaning it has two identical antigen binding sites. IgG is mainly synthesized in the secondary immune response to pathogens and can activate the classical complement pathway.

There are four IgG subclasses: IgG1, IgG2, IgG3 and IgG4. IgG1 accounts for approximately 65% of the total IgG. IgG2 represents a significant host defense against encapsulated bacteria. IgG is the only immunoglobulin that crosses the placenta because its Fc portion binds to receptors on the placental surface, protecting the newborn from infectious diseases. For this reason, IgG is the most abundant antibody in newborns (Wijdicks et al. 2017).

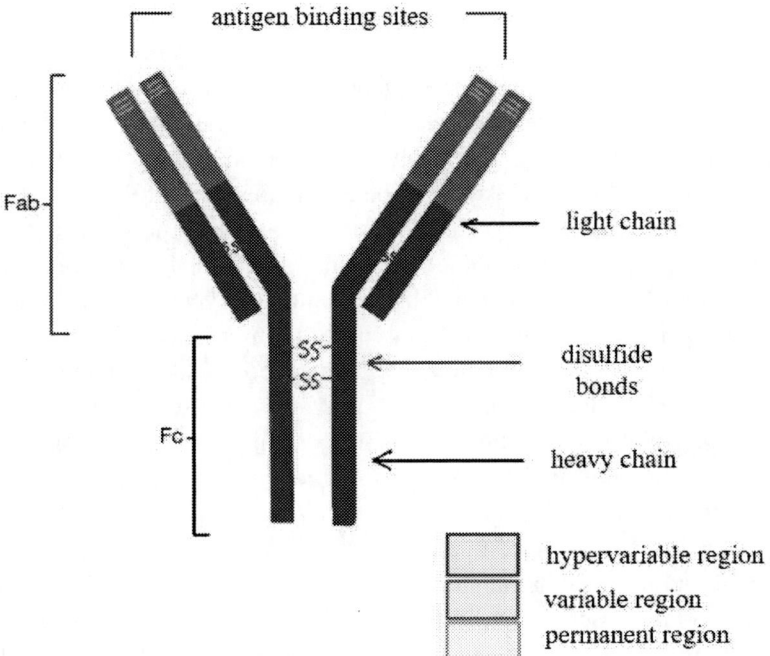

Figure 1. Schematic structure of immunoglobulins [Own elaboration].

IgA exists in two different molecular structures: monomeric (serum) and dimeric (secretory). Serum IgA has a molecular weight of 160 kD and a serum concentration of 3 mg/ml.

The secretory IgA (sIgA) has a molecular weight of 385 Kd and a mean serum concentration of 0.05 mg/ml. IgA is the main antibody contained in secretions such as saliva, tears, colostrum, and intestinal, genital and respiratory secretions. IgA has a secretory component that prevents enzymatic digestion. It activates an alternative pathway of complement activation.

IgE is a monomer with a molecular weight of 188 kD and a serum concentration of 0.00005 mg/ml. It protects the body against parasites and also binds to mast cell and basophil receptors, causing an allergic reaction.

IgD is a monomer with a molecular weight of 184 kD and is found to be low in serum (0.03 mg/ml). Its function against pathogens is as yet unknown. IgD is considered the BCR and plays an essential role in lymphocyte differentiation (Edlow et al. 2020).

The immune system is able to respond to many antigens thanks to the huge variety of immunoglobulins produced by plasma cells. Immunoglobulin light chains are encoded by V and J gene segments, and heavy chains are additionally encoded by D gene segments. Light chain coding is the accidental approximation of the V gene to the J gene, removing a DNA fragment lying between them during B-cell differentiation. As a result, the VJC complex is formed which is transcribed into an mRNA molecule. Finally, the introns and the J gene are cut. As for the coding of heavy chains, it is similar to the coding of light chains, but with the additional participation of the D gene. One of several dozen D genes is combined with one of several J genes, resulting in the formation of a DJ sequence that binds to one of several hundred V genes.

Purified immunoglobulin can be used to treat many immunological diseases, including antibody deficiencies, severe combined immune disorders (SCID), multiple sclerosis, myasthenia gravis, Kawasaki disease, systemic lupus erythematosus (SLE), organ transplants (Wijdicks et al. 2017).

Immunoglobulins are glycoproteins secreted by plasma cells in response to the presence of an antigen, i.e., a substance that is recognized as foreign by the host and that it specifically binds to it. There are 5 types of immunoglobulins in the human body: IgM, IgG, IgA, IgE, and IgD, which are characterized by structural and functional similarity, and the division results from the differentiation of the first domains of the heavy or light chain. Each type is made up of four polypeptide chains: two light and two heavy, which are linked by a disulfide bridge. Immunoglobulin light chains are encoded by V and J gene segments, and heavy chains additionally by D gene segments.

IgM is produced during the primary immune response and can activate the complement system. IgG is the highest percentage of plasma antibodies and is synthesized in the secondary immune response. IgA is found mainly in secretions, e.g., in saliva, tears and genital secretions, and IgE plays a role in the body's allergic response and protection against parasites. IgDs are the least recognized type and constitute the smallest amount in the blood serum.

Immunoglobulins play an essential role in the body's defense against pathogens, and thanks to their specificity and flexibility, they are an attractive option for the development of therapy against a wide range of diseases, including antibody deficiencies, severe combined immune disorders (SCID), multiple sclerosis, myasthenia gravis, Kawasaki disease, systemic lupus erythematosus (SLE), organ transplants.

References

Basyal B, KC P. Autoimmune Pancreatitis. 2021 Jul 19. In: StatPearls [Internet]. *Treasure Island* (FL): StatPearls Publishing; 2021 Jan–. PMID: 32809604.

Bracken SJ, Abraham S, MacLeod AS. Autoimmune Theories of Chronic Spontaneous Urticaria. *Front Immunol.* 2019 Mar 29;10:627.

Edlow AG, Li JZ, Collier AY, Atyeo C, James KE, Boatin AA, Gray KJ, Bordt EA, Shook LL, Yonker LM, Fasano A, Diouf K, Croul N, Devane S, Yockey LJ, Lima R, Shui J, Matute JD, Lerou PH, Akinwunmi BO, Schmidt A, Feldman J, Hauser BM, Caradonna TM, De la Flor D, D'Avino P, Regan J, Corry H, Coxen K, Fajnzylber J, Pepin D, Seaman MS, Barouch DH, Walker BD, Yu XG, Kaimal AJ, Roberts DJ, Alter G. Assessment of Maternal and Neonatal SARS-CoV-2 Viral Load, Transplacental Antibody Transfer, and Placental Pathology in Pregnancies During the COVID-19 Pandemic. *JAMA Netw Open.* 2020 Dec 1;3(12):e2030455.

Lu L, Zhang H, Zhan M, Jiang J, Yin H, Dauphars DJ, Li SY, Li Y, He YW. Antibody response and therapy in COVID-19 patients: what can be learned for vaccine development? *Sci China Life Sci.* 2020 Dec;63(12):1833-1849.

Wijdicks EF, Klein CJ. Guillain-Barré Syndrome. *Mayo Clin Proc.* 2017 Mar;92(3):467-479.

Chapter 13

SGLT1 in Head and Neck Cancers

Lidia Bieniasz, Dorota Bartusik-Aebisher[*] and David Aebisher
Medical College of The University of Rzeszów, Rzeszów, Poland

Abstract

SGLT1 and GLUT1 are involved in various glucose transport mechanisms to maintain glucose homeostasis in the body. In addition to glucose, SGLT1 protein also has a high affinity for galactose, therefore it plays an important role in the absorption of the small intestine. The tumor cell response to increased glucose requirements, in particular, increases the expression of membrane glucose transporters. Expression of SGLT and GLUT proteins in normal cells is characterized by high tissue specificity. Based on literature data in squamous cell carcinoma tissues of tongue cancer patients, immunohistochemistry showed increased SGLT1 expression and a high correlation between SGLT1 and EGFR expression. The discoveries and understanding of SGLT in neoplastic cells to date may help in the development of new anticancer therapies using SGLT1 or SGLT2 inhibitors already used in the treatment of diabetes.

Keywords: C-terminal domain SGLT1, head and neck cancers, epidermal growth factor receptor (EGFR), metastasize

---

[*] Corresponding Author's Email: dbartusikaebisher@ur.edu.pl.

In: The Biochemical Guide to Proteins
Editors: David Aebisher and Dorota Bartusik-Aebisher
ISBN: 979-8-88697-493-5
© 2023 Nova Science Publishers, Inc.

SGLT1 and GLUT1 are involved in various glucose transport mechanisms to maintain glucose homeostasis in the body. The GLUT1 protein is responsible for using the concentration gradient of the transport material to provide cells with a basic glucose supply by promoting transport. The SGLT1 cotransporter is the product of the SLC5A1 gene. In humans, it is a monomeric protein with a length of 664 amino acids and a molecular weight of 73 kDa. The secondary structure consists of 14 alpha-transmembrane helices, as in GLUT1, connected by loops extending above the cell membrane both inside the cell and in the extracellular space. The N-terminus of the SGLT1 protein extends beyond the cell, and the C-terminus is anchored to the cell membrane, forming the 14th terminal helix. There is an N-linked glycosylation site in the loop joining helices VI and VII, but the SGLT1 protein does not need to modify to become active (Wright et al., 2011).

There are two sugar binding sites in the C-terminal domain – one outside the cell and one inside the cell. The C-terminal domain, including the last five transmembrane helices, is also directly involved in the process of sugar transfer across the cell membrane. The sodium-glucose co-transporter SGLT1, found in the membranes of the epithelial cells of the small intestine and renal tubules, is in direct contact with the lumen of these tubules. It is responsible for the reabsorption of glucose from the small intestine and renal tubules, simultaneously transporting two sodium ion molecules and one glucose molecule in each cycle. It is characterized by high affinity to the substrate and low transmission efficiency. Glucose is transported to the cell against the concentration gradient, and the energy for this process comes from the energetically beneficial Na + ions flowing into the cell according to the concentration gradient and the membrane potential. This process is completely reversible and its direction depends only on the gradient of sodium and glucose concentrations (Wright et al., 1994).

SGLTs use a sodium transmembrane gradient as a driving force for glucose transport against the concentration gradient, which causes the concentration of glucose in the cell. The amino acid sequence of the SGLT1 protein also contains phosphorylation sites for protein kinases: 5 kinase C sites and 1 kinase A site, which play an important role in regulating the transport mechanism. In addition to glucose, SGLT1 protein also has a high affinity for galactose, therefore it plays an important role in the absorption of the small intestine. The tumor cell response to increased glucose requirements, in particular, increases the expression of membrane glucose transporters. Expression of SGLT and GLUT proteins in normal cells is characterized by high tissue specificity. This applies to the type and amount of a given transporter in a specific tissue. Neoplastic tissues not only overexpress these transporters, but also the presence of transport proteins that are not found in normal tissues at all (Ren et al., 2013).

Elevated levels of SGLT1 are found in colon cancer, prostate cancer, head and neck cancer, lung cancer, pancreatic cancer, and head and neck cancer. The presence of these proteins is closely related to the hypoxia of cells, especially in the tumor core. In most cases, overexpression of SGLT1 and GLUT1 proteins is associated with the tumor stage (tumor growth and reproduction, greater ability to metastasize) and poor prognosis. Due to the increase in the level of glucose transporter in cancer cells, they receive 20-30 times more glucose than normal cells. Little is known about the regulation of SGLT1 protein expression in cancer cells. One of the mechanisms leading to an increase in SGLT1 in tumors is its interaction with the epidermal growth factor receptor (EGFR). In epithelial tumors, which often overexpress EGFR, the level of SGLT1 protein increases due to EGFR stabilization, which satisfies the increased glucose demand of tumor cells, thus protecting them against autophagic cell death (Helmke et al., 2004).

Based on literature data (Helmke et al., 2004), SGLT1 protein was detected in 27 of 30 HNSCC tissues with heterogeneous staining limited to differentiated tumor cells. The differences between SGLT1 expression in cell culture and tissues may be the fact that less differentiated tumor cells proliferate in vitro (Madunić et al., 2018).

All cancer cells, without exception, require large amounts of glucose for uncontrolled proliferation. Based on literature data in squamous cell carcinoma tissues of tongue cancer patients, immunohistochemistry showed increased SGLT1 expression and a high correlation between SGLT1 and EGFR expression. At the moment, these studies are the only studies that have analyzed SGLT1 expression in these types of cancer, and moreover, studies on its expression and function in squamous epithelium are needed to explain the potential use of SGLT1 as a prognostic marker in head, neck and mouth cancers. The discoveries and understanding of SGLT in neoplastic cells to date may help in the development of new anticancer therapies using SGLT1 or SGLT2 inhibitors already used in the treatment of diabetes.

References

Helmke B M, Reisser C, Idzkoe M, Dyckhoff G, Herold-Mende C. Expression of SGLT-1 in preneoplastic and neoplastic lesions of the head and neck. *Oral Oncol.* 2004;40:28-35.

Madunić I V, Madunić J, Breljak D, Karaica D, Sabolić I. Sodium-glucose cotransporters: new targets of cancer therapy? *Arh. Hig. Rada Toksikol.* 2018 Dec 1;69(4):278-285.

Ren J, Bollu L R, Su F, Gao G, Xu L, Huang W C, Hung M C, Weihua Z. EGFR-SGLT1 interaction does not respond to EGFR modulators, but inhibition of SGLT1 sensitizes prostate cancer cells to EGFR tyrosine kinase inhibitors. *Prostate.* 2013 Sep;73(13):1453-61.

Wright E, Loo D, Hirayama B. Biology of human sodium glucose transporters. Physiol. Rev. 2011;91:733-94.

Wright E M, Loo D D, Panayotova-Heiermann M, Lostao M P, Hirayama B H, Mackenzie B, Boorer K, Zampighi G. "Active" sugar transport in eukaryotes. *J. Exp. Biol.* 1994;196:197-212.

Chapter 14

Major Histocompatibility Antigens

Klaudia Fikas, Dorota Bartusik-Aebisher[*] and David Aebisher
Medical College of The University of Rzeszów, Rzeszów, Poland

Abstract

Primary histocompatibility antigens provide immune protection to organisms that express the genes that synthesize these glycoproteins. Understanding the molecules of the main histocompatibility complex has made it possible to demonstrate once again in the world of science the importance of proteins in the body. MHC, which by their diversity play an enormous role in medicine, are also significant in maintaining individual identity. Natural selection is often found to be insufficient to keep MHC gene polymorphisms constant so as to outweigh the natural randomized allele loss process. Research on the spread of diseases (not only infectious) in immunocompromised animal populations may form the basis for further tests on the control of human pathogens in correlation with MHC proteins.

Keywords: major histocompatibility complex (MHC), immune system, human leukocyte antigens (HLA), histocompatibility

The MHC (major histocompatibility complex) system is a specific structure responsible for the selection of antigens in the organism. The major histocompatibility system identifies potentially harmful molecules, including

---

[*] Corresponding Author's Email: dbartusikaebisher@ur.edu.pl.

In: The Biochemical Guide to Proteins
Editors: David Aebisher and Dorota Bartusik-Aebisher
ISBN: 979-8-88697-493-5
© 2023 Nova Science Publishers, Inc.

it induces an immune humoral response, consisting mainly of the synthesis of antibodies by B lymphocytes. Efficient functioning of this system determines the protection of the organism in terms of intergality and allows for the preservation of a specific distinction between an individual and others, regardless of whether he belongs to a given species. In the human system, we assume the term human leukocyte antigens – HLA (Bakela et al. 2018). The genes encoding the histocompatibility proteins are located on chromosome 6. Due to the different functions, the HLA classes have been distinguished:

- MHC class I
- MHC class II

Class III MHCs are also known, however, their structure and role in the system do not allow them to be included among HLA. Class I molecules are present on all cells containing the nucleus, while class II molecules on antigen-presenting cells – as glycoproteins are bound to the cell membrane. The described proteins constitute an obstacle in the field of transplantology, as they are responsible for a more or less intense immune response and may cause rejection of the transplanted tissue or organ. Knowledge about MHC was expanded only in the second half of the twentieth century, so a more comprehensive understanding of these proteins may be a milestone towards a breakthrough in the field of transplantation (Godfrey et al. 2019).

The immune system is made up of a set of mechanisms whose common goal is to recognize foreign proteins and trigger a cascade of reactions known as the immune response. It is the consequences of the actions of this system that constitute the basis for the survival of an organism highly situated in the taxonomic hierarchy. Thanks to the synthesis of specific antibodies, we are able to fight invisible and inconspicuous bacteria and viruses. The individually diverse packages of produced antibodies also determine to some extent which individuals within a species will pass their genes on to their progeny.

The oldest descriptions of the system date back to 1936, when the British immunologist Peter Gorer and geneticist George Snellle discovered H-2, the 2nd MHC locus in mice, which later turned out to have analogous functions to human leukocyte antigens. In addition, the relationship of proteins synthesized on the basis of these genes with the problem of transplant rejection has been demonstrated. In 1980, three researchers were awarded the Nobel Prize in Physiology or Medicine for research into the genetically determined structures on the surface of cells that regulate immune responses. It was not until 1999 that the sequence of the major histocompatibility complex proteins was fully described and published in Nature thanks to teams of scientists from the United Kingdom, the United States and Japan (Godfrey et al. 2018).

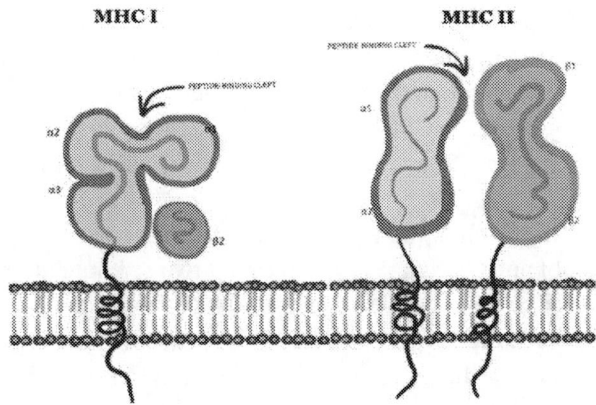

Figure 1. Diagram of the main particle structure of class I and II molecules. MHC I consists of two chains, of which α (or more precisely - the $\alpha 3$ subunit) anchors in the cell membrane. The chain β is a separate unit here. Similarly, MHC class II consists of two chains, however both α and β bind to the phospholipid bilayer. In MHC I, two subunits - $\alpha 1$ and $\alpha 2$ create a characteristic groove for binding with short peptides. In MHC II, the $\alpha 1$ subunit forms a groove together with the $\beta 1$ subunit [Own study].

In humans, leukocyte antigens and elements of the complement system comprise a group of over one hundred genes in the area of about four million base pairs that are located on the short arm of somatic chromosome 6. The major compatibility complex genes show polymorphism, that is, they occur in more than one version in populations. More than a thousand alleles (variants) of these genes have been discovered in humans. So it is safe to say that they are an extreme case of polymorphism that can be explained as a specific kind of natural selection.

Mutation in the case of most proteins means the risk of disturbing the functioning of the body and often – its death. The frequently occurring amino acid substitutions should therefore persist much less frequently compared to silent mutations. However, the opposite was noted for MHC proteins. This is explained as follows: mutations, although not the best variants in terms of evolution, persist because their multiplicity within a species is able to give higher organisms an advantage over potentially pathogenic pathogens, which then are not able to threaten all types of proteins MHC present in the population (Kamiya et al. 2020).

Taking into account the functions of the proteins of the main histocompatibility complex, i.e., the binding of pathogens and their subsequent presentation to T lymphocytes, research began on the relationship between the occurrence of hundreds (perhaps thousands) of MHC alleles with the constant presence of many diseases in the human and animal environment. Among the subjects of interest, the following diseases are distinguished in humans: immune thrombocytopenia, hemolytic anemia, and Hashimoto's disease. This relationship has been demonstrated, inter alia, for HLA class II in hepatitis C infection, but this is not the only publication confirmed by research.

To perform a transplant (apart from the team of doctors), two people are required: the donor and the recipient. As they are two separate units, they differ not only in their external appearance, but also in the genetic structure, specificity of other antigens (not related to the transplanted organ or tissue) – the so-called minor histocompatibility antigens), multiples of satellite sequences and many other factors. To safely carry out transplants, it is necessary to carefully test the donor organism and biora for human leukocyte antigens. Transplantation will be successful if the antigenic mismatch does not exceed 1/10 of the incompatible alleles, so without analyzing the MHC structure in the body, the transplantology would not be able to function (Wang et al. 2019).

Primary histocompatibility antigens provide immune protection to organisms that express the genes that synthesize these glycoproteins. When a given pathogen cannot be inactivated by the basic types of immunity (innate and non-specific immunity), an acquired (secondary) response is triggered. The humoral and cellular responses arise only after self-antigens (histocompatibility) are presented and distinguished from pathogens by appropriate groups of MHC cells to Th lymphocytes, i.e., to helper lymphocytes. Understanding the molecules of the main histocompatibility complex has made it possible to demonstrate once again in the world of science the importance of proteins in the body. MHC, which by their diversity play an enormous role in medicine, are also significant in maintaining individual identity. In addition, animal species threatened with extinction provide the starting point for gene research for the compatibility complex. In populations that are constantly declining, the pool of genetic variation is also diminishing. Natural selection is often found to be insufficient to keep MHC gene polymorphisms constant so as to outweigh the natural randomized allele loss process. Ultimately, a reduced gene pool in a population may lead to a reduction in the immune response capacity of an individual. Research on the spread of diseases (not only infectious) in immunocompromised animal populations may form the basis for further tests on the control of human pathogens in correlation with MHC proteins.

References

Bakela K, Athanassakis I. Soluble major histocompatibility complex molecules in immune regulation: highlighting class II antigens. *Immunology.* 2018 Mar;153(3):315-324.

Godfrey DI, Koay HF, McCluskey J, Gherardin NA. The biology and functional importance of MAIT cells. *Nat Immunol*. 2019 Sep;20(9):1110-1128.

Godfrey DI, Le Nours J, Andrews DM, Uldrich AP, Rossjohn J. Unconventional T Cell Targets for Cancer Immunotherapy. *Immunity.* 2018 Mar 20;48(3):453-473.

Kamiya J, Kang W, Yoshida K, Takagi R, Kanai S, Hanai M, Nakamura A, Yamada M, Miyamoto Y, Miyado M, Kuroki Y, Hayashi Y, Umezawa A, Kawano N, Miyado K. Suppression of Non-Random Fertilization by MHC Class I Antigens. *Int J Mol Sci.* 2020 Nov 19;21(22):8731.

Wang J, Sanmamed MF, Datar I, Su TT, Ji L, Sun J, Chen L, Chen Y, Zhu G, Yin W, Zheng L, Zhou T, Badri T, Yao S, Zhu S, Boto A, Sznol M, Melero I, Vignali DAA, Schalper K, Chen L. Fibrinogen-like Protein 1 Is a Major Immune Inhibitory Ligand of LAG-3. *Cell.* 2019 Jan 10;176(1-2):334-347.e12.

Chapter 15

Vascular Endothelial Growth Factor (VEGF)

Eliza Gryboś, Dorota Bartusik-Aebisher[*] and David Aebisher
Medical College of The University of Rzeszów, Rzeszów, Poland

Abstract

Vascular endothelial growth factor (VEGF) is a formation of new blood vessels, angiogenesis – i.e., the creation of branches from already existing vessels. The endothelial growth factor also plays a very important role in the process of wound healing, the course of pregnancy in the construction of the placenta and in the menstrual cycle in women – in the regeneration of the endometrium. In other words, VEGF is necessary for the growth and survival of the vascular endothelium. Its increased activity is observed in conditions such as cancer, diabetic retinopathy, rheumatoid arthritis, psoriasis, and in bronchitis and the clinical course of asthma. As shown by many studies, VEGF plays an important role in the progression of neoplastic diseases, but it is a poor prognostic factor. It is responsible for the development and activation of metastasis formation of many types of solid tumors. The conducted studies allow to conclude that the reduction of VEGF secretion is important in inhibiting the angiogenesis process. Therefore, the introduction of anti-angiogenic therapy in the treatment of neoplastic diseases gives very promising results.

Keywords: vascular endothelial growth factor (VEGF), bronchitis, anti-angiogenic, pro-angiogenic, neoplastic diseases

---

[*] Corresponding Author's Email: dbartusikaebisher@ur.edu.pl.

In: The Biochemical Guide to Proteins
Editors: David Aebisher and Dorota Bartusik-Aebisher
ISBN: 979-8-88697-493-5
© 2023 Nova Science Publishers, Inc.

Vascular endothelial growth factor (VEGF) began to gain more and more interest from scientists only since 1995. When more and more scientific articles about this protein appeared. Initially, VEGF was described as a factor that stimulates the permeability of blood vessels, but is now considered one of the main factors regulating the angiogenesis process. VEGF acts on the human body already at the stage of embryogenesis, where it stimulates the development of a normal network of blood vessels, and at the stage of further development, it is responsible for vasculogenesis – i.e., the formation of new blood vessels, angiogenesis – i.e., the creation of branches from already existing vessels (Apte et al. 2019). The endothelial growth factor also plays a very important role in the process of wound healing, the course of pregnancy in the construction of the placenta and in the menstrual cycle in women – in the regeneration of the endometrium. In other words, VEGF is necessary for the growth and survival of the vascular endothelium. This factor, apart from participating in normal angiogenesis, is also involved in the pathological one. Its increased activity is observed in conditions such as cancer, diabetic retinopathy, rheumatoid arthritis, psoriasis, and in bronchitis and the clinical course of asthma (Apte et al. 2019).

Purified VEGF protein was first obtained in 1989. It was done by D. Gospodarowicz, J. Abraham, J. Schilling and N. Ferrara, W. J. Henzel. VEGF genes have been identified in the last few years, and this makes VEGF the first vascular growth factor described in this way in the entire family of these factors. The genes of this protein, i.e., VEGF-A, -B, -C, -D, are located in the genome of mammalian cells, and in humans the exact location of the gene encoding the VEGF growth factor is chromosome 6p21.3. Its structure consists of eight exons separated by seven non-coding regions. VEGF genes result from alternative mRNA splicing. This causes VEGF to exist in several isomorphic forms. This was demonstrated by cDNA sequence analyzes. The isomorphic forms differ in the length of the amino acid chain, and thus also in their properties. The process of alternate splicing of the VEGF gene produces several isoforms of the same protein. Variants containing 121, 145, 148, 162, 165, 183, 189 and 206 amino acids, respectively, are known. There is also placental growth factor (PIGF) which exists as two isoforms PIGF131 and PIGF152. It is identified as the first factor homologous to VEGF (Karaman et al. 2018).

VEGF plays a key role in neoplastic progression due to the fact that it is the main factor responsible for the angiogenesis process. Angiogenesis (neovascularization) is a multistage process regulated by stimulants (pro-angiogenic) and inhibitory (anti-angiogenic) factors. These factors are responsible for physiological and pathological angiogenesis. Disturbances in the balance between pro-angiogenic and anti-angiogenic factors result in the predominance of activating (pro-angiogenic) factors, and thus the process of excessive angiogenesis (Karaman et al. 2018).

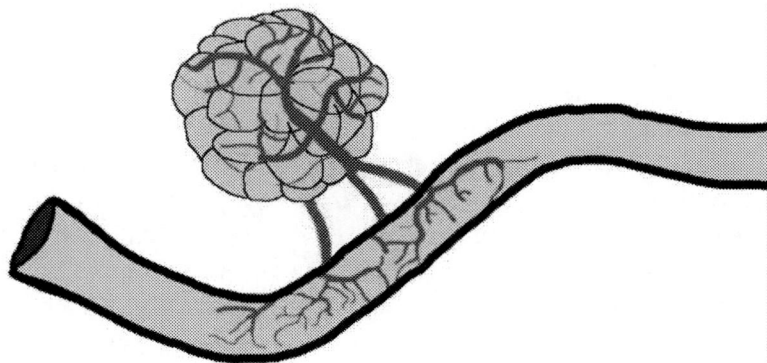

Figure 1. Pathological angiogenesis.

The sequence of such events promotes the development of a neoplastic tumor. An additional and the most important factor enhancing the process of pathological angiogenesis as well as enhancing the secretion of VEGF is hypoxia (hypoxia) (Melincovici et al. 2018). The hypoxic microenvironment is inside the tumor. On the basis of this, it can be easily concluded that VEGF is a poor prognostic factor in neoplastic disease. It is responsible for the activation and metastasis of many solid tumors. The formation of new blood vessels in a tumor is necessary for the growth of a so-called primary tumor and for the formation of tumor metastasis. Without the presence of new blood vessels, and therefore with a poor oxygen supply, tumor growth of more than 1-2 mm3 would not be possible (Melincovici et al. 2018). The next stages of carcinogenesis (further tumor growth) cannot take place without additional blood supply to cells, therefore the phenomenon of angiogenic switching occurs, i.e., the tumor cells acquire an angiogenic phenotype. This means that cancer cells move from a proliferative (increasing the number of cells) phenotype to an angiogenic phenotype. Then, irreversible genetic modifications occur in the genome of neoplastic cells, which results in an increased, uncontrolled production of angiogenic factors such as VEGF. The advantage of angiogenic factors over anti-angiogenic factors causes the formation of a new capillary network inside the tumor (Siveen et al. 2017). Better blood supply to cancer cells, and therefore nutrients and oxygen, allows for its better growth. However, newly formed vessels do not have the typical structure of normal vessels because they are characterized by abnormal irregular shape, morphological immaturity and increased permeability. These features result in the fact that, despite the large number of blood vessels, the tumor cells are not supplied with the appropriate amount of oxygen, resulting in tissue hypoxia, which in turn causes the activation of angiogenic factors, especially VEGF. Hypoxia plays a very important role in this process, and as a result of it, the HIF 1 alpha protein (Hypoxia Inducible Factor) is released. This protein stimulates the expression of genes encoding proteins involved in the processes of glycolysis and angiogenesis. As a result, cancer cells are able to adapt to the low oxygen concentration in the environment. HIF 1 protein works by increasing the synthesis and release of VEGF. VEGF then binds to appropriate VEGFR receptors on the surface of endothelial cells, which activates the tyrosine kinase (Src kinase) pathway and leads to angiogenesis (Siveen et al. 2017).

As shown by many studies, VEGF plays an important role in the progression of neoplastic diseases, but it is a poor prognostic factor. It is responsible for the development and activation of metastasis formation of many types of solid tumors. It should also be emphasized that a high level of VEGF may indicate angiogenic switching, and thus it may be an early stage of metastasis formation. The experiment carried out on two cell lines (colorectal cancer cells) with different degrees of malignancy proved that the level of VEGF secretion is directly proportional to the malignancy level of colorectal cancer cells. (Viallard et al. 2017) Other studies also confirm that VEGF can activate the formation of the Bcl protein – it protects cancer cells against apoptosis, which additionally adversely affects the effectiveness of the therapy and worsens the prognosis in cancer. The conducted studies allow to conclude that the reduction of VEGF secretion is important in inhibiting the angiogenesis process. Therefore, the introduction of anti-angiogenic therapy in the treatment of neoplastic diseases gives very promising results. For this purpose, inhibitors of angiogenic factors – VEGF inhibitors, antibodies blocking receptors located directly in tumor cells, and tyrosine kinase inhibitors that inhibit the transmission of pro-angiogenic signals are used (Viallard et al. 2017).

References

Apte RS, Chen DS, Ferrara N. VEGF in Signaling and Disease: Beyond Discovery and Development. *Cell*. 2019;176(6):1248-1264.

Karaman S, Leppänen VM, Alitalo K. Vascular endothelial growth factor signaling in development and disease. *Development*. 2018;145(14):dev151019.

Melincovici CS, Boşca AB, Şuşman S, Mărginean M, Mihu C, Istrate M, Moldovan IM, Roman AL, Mihu CM. Vascular endothelial growth factor (VEGF) - key factor in normal and pathological angiogenesis. *Rom J Morphol Embryol*. 2018;59(2):455-467. PMID: 30173249.

Siveen KS, Prabhu K, Krishnankutty R, Kuttikrishnan S, Tsakou M, Alali FQ, Dermime S, Mohammad RM, Uddin S. Vascular Endothelial Growth Factor (VEGF) Signaling in Tumour Vascularization: Potential and Challenges. *Curr Vasc Pharmacol*. 2017;15(4):339-351.

Viallard C, Larrivée B. Tumor angiogenesis and vascular normalization: alternative therapeutic targets. *Angiogenesis*. 2017;20(4):409-426.

Chapter 16

T Cell Receptor

Paulina Nowak, Dorota Bartusik-Aebisher* and David Aebisher

Medical College of The University of Rzeszów, Rzeszów, Poland

Abstract

The receptors on T cells are fairly simple and ensure the safety of our body and protection against all unwanted microorganisms that can disrupt its work and pose a real danger. Thanks to them, the body knows what its component is and what is an intruder and must be eliminated. Also, thanks to the efforts of people involved in genetic aspects activities, new therapies are developed or therapies are discovered, treatment methods used mainly in the treatment of cancer. Taking into account that neoplastic diseases are among the diseases of civilization, the hopes resulting from the possibility of using T-cell receptors are an opportunity for many people for whom there is currently no rescue.

Keywords: T cell receptors (TCR), complementarity determining regions (CDRs), lymphocytes, autoimmunity

T cell receptors are made up of protein chains. They appear on the surface of T lymphocytes. T cell receptors have 2 protein chains in their structure. They are involved in creating the site that is involved in attaching antigens. In 95% of cases, these receptors are composed of alpha and beta chains, but in 5% of cases of gamma and delta chains. The chains are connected by disulfide

---

* Corresponding Author's Email: dbartusikaebisher@ur.edu.pl.

In: The Biochemical Guide to Proteins
Editors: David Aebisher and Dorota Bartusik-Aebisher
ISBN: 979-8-88697-493-5
© 2023 Nova Science Publishers, Inc.

bridges, whose task is to stabilize their structure. The most important structural feature of T-cell receptors is the presence of a specific site specialized for attaching specific antigens. They thus ensure the specificity of their binding. T cell receptors are an integral part of the T cell membrane. (Figure 1) They are associated with the CD3 co-receptor. These molecules, in turn, are necessary to convert antigen binding outside the cell into intracellular signals (Alcover et al. 2018).

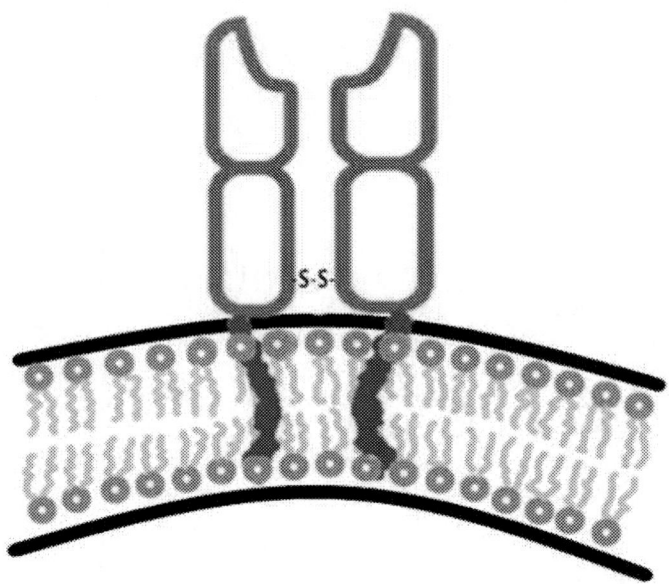

Figure 1. Diagram of the structure of the T-cell receptor [Own elaboration].

MHC molecules, the major histocompatibility complex, have a stimulating effect on TCR. The signal received by the described receptors causes a number of changes in the T-lofcite cell. They are needed for the cell to be able to undergo further differentiation and perform its functions properly. The protein complex also participates in stimulating T lymphocytes to respond to the attached antigen. The genes responsible for this take part in the development of receptors, and they themselves are shaped and mature during the processes taking place in the thymus. Due to specific genetic mechanisms, they are an important starting point used in therapies. There is now a lot of interest in these structures. They also play many important roles in our body and are involved in many regulatory processes, so they can be used in various ways (Alcover et al. 2018).

These proteins, apart from immunoglobulins, are the only antigen-binding molecules. However, they differ from immunoglobulins in that they bind only to antigens, which are fragments of proteins and are attached to MHC molecules, which, in turn, are found in the cell membrane. Delta gamma lymphocytes also have the ability to recognize antigens that are linked to CD1 molecules. T cell receptors have two chains. There is always a variable part and a constant part in the chain. It is a fragment of the lymphocyte cell because it is connected to it by two sections of the endothelial and the intracellular. There are 'three hypervariable regions or complementarity determining regions (CDRs)' in the variable portions of the chains (Chandran et al. 2019).

The genetic information about these chains is encoded in our genome on chromosomes 7 and 14. Variable parts of alpha and gamma chains are encoded by V and J gene segments, and those related to delta and beta chains in V, D and J gene segments. Recombination of genes concerning TCR and their synthesis takes place in the thymus. However, recombination of these genes specifically related to the gamma chain may start in thymocytes. Initially, TCR receptors consisting of the B-chain and pre-alpha appear in the thymus. This is how the original TCR is made. It stimulates the proliferation and further stages of T-cell precursor maturation. Sometimes a phenomenon called receptor editing occurs. It is based on the recombination of TCR genes, thanks to which the lymphocyte that was to be negatively selected has a chance of survival. Protection against autoimmunity of T lymphocytes is the effect of low frequency of somatic mutations in TCR genes (Chandran et al. 2019).

Genetic engineering has achieved a great deal with the modification of T-cell receptors. There are now possibilities of activating or modifying genes responsible for recognizing cells with a specific antigen, e.g., in the case of cancer cells. Also those that do not have MHC particles. Due to these discoveries, the subject of these structures appears quite often in medicine. Precisely in the context of their use in the treatment of cancer. These diseases are very dangerous and take a heavy toll on people, including young people and children. Receptors and their use give high hopes and chances of recovery. The most famous treatment in which they play a major role is a relatively new method, i.e., chimeric antigent receptors, abbreviated as CAR-t cell receptor. It is a type of gene therapy which consists in modifying genes of T cells. Modification involves adapting these cells to recognize antigens of appropriate tumor cells in order to destroy them later. Due to the modification of appropriate genes, the receptors of T cells, which are responsible for the recognition of a pathologically altered cell, are changed (Mariuzza et al. 2020).

Due to their abilities and specificity, these receptors can perform tasks and take part in the destruction of neoplastic cells. CAR-t therapy can currently be used in the treatment of B-cell acute lymphoblastic leukemia, B-cell malignant lymphomas and multiple myeloma. Its effectiveness is at a high level, which means high hopes for the entire medical community. The second method with the use of T lymphocyte receptors, i.e., TCR-t, is also used. The difference between CAR-T and TCR-T therapy is that the antigens they bind to are different, since those with TCR can only bind to MHC proteins. TCR therapy is now also a more versatile method. Therefore, it can be applied to a much larger number of neoplastic diseases. However, both types of these therapies currently face very similar limitations and are therefore not widely used. However, they are still being developed (McLellan et al. 2019).

The receptors on T cells are fairly simple in structure. However, it is very important as it conditions the performance of all functions. Due to the presence of specific receptors on the surface, lymphocytes can also perform their tasks properly. It is these functions that determine and ensure the safety of our body and protection against all unwanted microorganisms that can disrupt its work and pose a real danger. Also thanks to them, the body knows what is its component and what is an intruder and must be eliminated. That is why they are very important and our existence would not take place without them. The receptors, performing their respective tasks, are the second cells, apart from immunoglobulins, that can bind to antigens. The binding specificity of these antigens is what sets these proteins apart and defines them. This fact and numerous information and data related to genetic aspects constitute a very important part of research and what genetic engineering can do. Also, thanks to the efforts of people involved in these activities, new therapies are developed or therapies are discovered, treatment methods used mainly in the treatment of cancer. Taking into account that neoplastic diseases are among the diseases of civilization, the hopes resulting from the possibility of using T-cell receptors are an opportunity for many people for whom there is currently no rescue. Thanks to this, it can be concluded that the described proteins are irreplaceable, they are the subject of many studies and the interests of scientists. It follows that an inconspicuous molecule can be something that allows people to recover from serious illnesses and the body to maintain a physiological state.

References

Alcover A, Alarcón B, Di Bartolo V. Cell Biology of T Cell Receptor Expression and Regulation. *Annu Rev Immunol.* 2018 Apr 26;36:103-125.

Biernacki MA, Brault M, Bleakley M. T-Cell Receptor-Based Immunotherapy for Hematologic Malignancies. *Cancer J.* 2019 May/Jun;25(3):179-190.

Chandran SS, Klebanoff CA. T cell receptor-based cancer immunotherapy: Emerging efficacy and pathways of resistance. *Immunol Rev.* 2019 Jul;290(1):127-147.

Mariuzza RA, Agnihotri P, Orban J. The structural basis of T-cell receptor (TCR) activation: An enduring enigma. *J Biol Chem.* 2020 Jan 24;295(4):914-925.

McLellan AD, Ali Hosseini Rad SM. Chimeric antigen receptor T cell persistence and memory cell formation. *Immunol Cell Biol.* 2019 Aug;97(7):664-674.

Chapter 17

Serum Albumin

Kamil Jugo, Dorota Bartusik-Aebisher[*] and David Aebisher

Medical College of The University of Rzeszów, Rzeszów, Poland

Abstract

> Albumin is a low molecular weight, globular protein that is one of the main components of human blood plasma. Its synthesis takes place in liver cells – hepatocytes. Apart from regulating oncotic pressure, it has many important functions, influencing the functioning of the whole organism. Albumin has been shown to have anticoagulant activity by indirectly neutralizing factor Xa and preventing platelet aggregation, and is therefore involved in the maintenance of haemostasis. Low levels of albumin may indicate liver diseases such as cirrhosis and viral hepatitis (impaired protein synthesis), in kidney diseases, especially in nephrotic syndrome. The decreased concentration of albumin may be states of malabsorption or deficiency of nutrients. Therefore, albumin is a very important diagnostic indicator presenting the patient's health condition.

> **Keywords:** serum albumin, hypoalbuminemia, sulfenic acid, Leśniewski-Crohn's disease, ulcerative colitis, liver cells

Plasma albumin may also be otherwise called blood albumin. It is a globular (ball-shaped) protein with a low molecular weight (66.5 kDa). Its molecule is made up of 585 amino acids arranged in a single chain. This chain is organized

---

[*] Corresponding Author's Email: dbartusikaebisher@ur.edu.pl.

In: The Biochemical Guide to Proteins
Editors: David Aebisher and Dorota Bartusik-Aebisher
ISBN: 979-8-88697-493-5
© 2023 Nova Science Publishers, Inc.

into three repeating homologous domains (I, II, III), each of which is additionally divided into subdomains A and B. There are no prosthetic groups or carbohydrates in the structure of the amino acid chain of albumin. The albumin coding gene has a location in humans on chromosome 4 of locus 4q13.3. The serum albumin concentration is normal if it is in the range of 35-50 g/l. The determination of its concentration in plasma is an important indicator of the patient's health due to its numerous functions in the body, and it can be used in clinical practice due to its low cost and low risk of an erroneous result. This protein is synthesized in hepatocytes (Fig. 1), so too low its concentration may indicate liver diseases, but it is not a specific test, as the correct concentration of albumin may occur in pathological conditions of this organ. Albumin, despite being smaller than globulin, maintains as much as 80% of blood plasma oncotic pressure, and is also a transmitter for antithrombin and for the heparin cofactor, so it plays a role in preventing blood clotting. The amount of albumin in the plasma decreases when there is inflammation in the body (Arques et al., 2018).

The serum albumin concentration is influenced by the rate of its synthesis and catabolism of this protein, the distribution of the protein outside the vessels and its exogenous loss. A possible use of serum albumin determination in predicting the risk of many cardiovascular diseases has recently been discovered. Based on epidemiological data, it can be concluded that a decreased serum albumin level is associated with the occurrence of ischemic heart disease, stroke, atrial fibrillation and venous thromboembolism, without being associated with other risk factors for these diseases. Hypoalbuminemia has also occurred in many diseases, such as stroke, congenital heart disease, myocardial infarction, coronary artery disease, and infective endocarditis, despite the presence of standard risk factors. The association of low albumin concentration with the occurrence of coronary artery disease may be related to its indirect function of regulating cholesterol transport. Albumin does not have the proper binding sites for cholesterol, but nevertheless stimulates its transport by increasing water diffusion, which increases the outflow of cholesterol from the cells. In addition, albumin binds specifically to cholesterol transport vesicles (Litus et al., 2018).

Studies have also shown the effect of albumin on blood clotting, but its influence on this process is inconclusive. The anticoagulant activity of albumin has been demonstrated by binding antithrombin and indirectly thereby neutralizing blood coagulation factor Xa. In addition, albumin binds to epinephrine, and coats the surfaces of the vascular membrane, thereby preventing platelet aggregation. Another anticoagulant mechanism of albumin is the intensification of the synthesis of nitric oxide (II) in macrophages, which is a strong factor inhibiting platelets, as well as the binding of prostacyclin inhibitor (PGI2), which prevents its breakdown and increases the concentration of this compound with the increase in serum albumin concentration. Albumin, in contrast to the above mechanisms, exhibits a post-thrombotic effect by increasing the TF tissue factor in monocytes, which has consequences in the activation of the blood coagulation process. Overall, however, the anticoagulant properties of albumin predominate, so we can consider it a weak anticoagulant. Therefore, an albumin deficiency will increase the risk of venous thromboembolism (Nishi et al., 2020).

Albumin also has antioxidant properties. Oxidative stress caused by an excessive amount of free radicals causes the oxidation of LDL cholesterol, contributing to the formation of atherosclerotic plaques, as well as fatty acids and collagen of the skin contributing to its aging. It also contributes to the emergence of neurodegenerative diseases in the future. ROS reactive oxygen species and RNS reactive nitrogen species are mainly responsible for the occurrence of oxidative stress. Under physiological conditions, about 60% is in a reduced form called mercaptoalbumin, which has a free thiol group in the cys-34 residue. This free thiol group accounts for 80% of the reactive thiols in the plasma and has the ability to trap oxidants such as hydrogen peroxide, hypochlorous acid and peroxynitrite (Rabbani et al., 2019). Upon capture of the free radical, it is reduced and the cys-34 thiol group is oxidized to sulfenic acid, which can be further oxidized to sulfinic and sulfonic acids. Additionally, the indirect antioxidant property of albumin results from the binding of transition metal ions – iron and copper. Iron and copper ions have the ability to catalyze the oxidation of hydrogen peroxide to ROS, so binding them by albumin will prevent the formation of reactive oxygen species. Albumin is responsible for maintaining the plasma redox balance. Plasma albumin deficiency causes deformation of red blood cells. The abnormal shape of red blood cells increases the viscosity of blood circulating in the vessels, which increases the risk of a heart attack. Additionally, it maintains good capillary membrane permeability (Wang et al., 2018).

Albumin is a low molecular weight, globular protein that is one of the main components of human blood plasma. Its synthesis takes place in liver cells – hepatocytes. Its primary function is to maintain oncotic blood pressure. The determination of its concentration may be an important warning marker in cardiovascular diseases. Apart from regulating oncotic pressure, it has many important functions, influencing the functioning of the whole organism. It regulates the transport of many substances, incl. cholesterol, and is also a messenger for the heparin cofactor and antithrombin. Albumin has been shown to have anticoagulant activity by indirectly neutralizing factor Xa and preventing platelet aggregation, and is therefore involved in the maintenance of haemostasis. Its antioxidant properties are also important. It captures metal ions which are catalysts for the formation of free oxygen radicals, and also catches free radicals themselves, reducing them to less aggressive chemical compounds. It also ensures the correct shape of erythrocytes. Low levels of albumin may indicate liver diseases such as cirrhosis and viral hepatitis (impaired protein synthesis), in kidney diseases, especially in nephrotic syndrome. The reason for the decreased concentration of albumin may be states of malabsorption or deficiency of nutrients: low-protein diet, Leśniewski-Crohn's disease or ulcerative colitis. Increased albumin catabolism occurs in the event of blood loss, disseminated tumors or extensive burns. Increased albumin concentration is usually not a sign of serious pathological conditions and may be the result of a high-protein diet, dehydration or medications. Therefore, albumin is a very important diagnostic indicator presenting the patient's health condition.

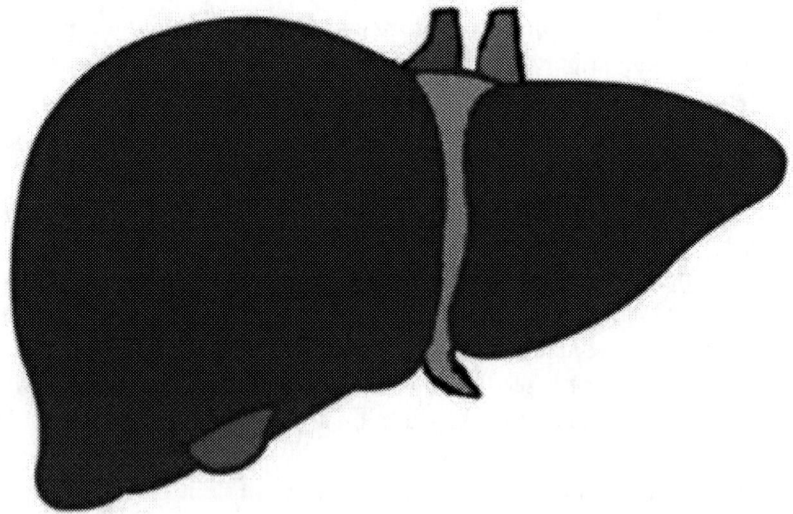

Figure 1. Liver-site of albumin synthesis [Own elaboration].

References

Arques S. Human serum albumin in cardiovascular diseases. *Eur. J. Intern. Med.* 2018; 52:8-12.

Litus E A, Permyakov S E, Uversky V N, Permyakov E A. Intrinsically Disordered Regions in Serum Albumin: What Are They For? *Cell Biochem. Biophys.* 2018;76(1-2):39-57.

Nishi K, Yamasaki K, Otagiri M. Serum Albumin, Lipid and Drug Binding. *Subcell Biochem.* 2020;94:383-397.

Rabbani G, Ahn S N. Structure, enzymatic activities, glycation and therapeutic potential of human serum albumin: A natural cargo. *Int. J. Biol. Macromol.* 2019;123:979-990.

Wang J, Zhang B. Bovine Serum Albumin as a Versatile Platform for Cancer Imaging and Therapy. *Curr. Med. Chem.* 2018;25(25):2938-2953.

Chapter 18

Ferritin

Kamil Chwaliszewski, Dorota Bartusik-Aebisher[*] and David Aebisher
Medical College of The University of Rzeszów, Rzeszów, Poland

Abstract

Ferritin is a protein necessary for the proper functioning of the entire body with a molecular weight of 440kDa. The much attention is paid to the role of mitochondrial ferritin. Ferritin is also a clinically important protein: it is responsible for the pathological processes in the body that lead to the development of diseases. Research on ferritinophagy indicates the presence of other, more intrusive, previously unknown, apoptotic mechanisms that enable the regulation and maintenance of homeostasis. A more detailed understanding of such mechanisms may be a place of action for potential chemotherapeutic agents or other active substances. The described phenomena make it possible to understand a number of processes aimed at providing iron ions for the synthesis of certain substances in the body; especially in the course of hematopoiesis. The most surprising fact is that despite the long history of research on this protein, the mechanism of the relationship between its concentration and the above-mentioned factors is still not fully understood.

Keywords: ferritin, hematopoiesis, malignant diseases, Alzheimer's disease, Parkinson's disease

---

[*] Corresponding Author's Email: dbartusikaebisher@ur.edu.pl.

In: The Biochemical Guide to Proteins
Editors: David Aebisher and Dorota Bartusik-Aebisher
ISBN: 979-8-88697-493-5
© 2023 Nova Science Publishers, Inc.

Ferritin is a protein with a molecular weight of 440kDa. It conditions the maintenance of homeostasis related to iron storage in the body. Iron is a component of human proteins: mainly hemoglobin and myoglobin, as well as a catalyst for the conversion of ribonucleotides to deoxyribonucleotides, as it is part of ribonucleotide reductase. Ferritin is also a clinically important protein: it is responsible for the pathological processes in the body that lead to the development of diseases. Its increased concentration in the body is often associated with the development of coronary artery diseases and malignant diseases (malignitas morbi) (Arosio et al. 2017). It is composed of a total of 24 subunits: 11 of the heavy type H (heavy) and 13 of the light type L (light). Type H shows ferroxidase activity which determines the attachment of iron ions to the protein. The L subunits are responsible for the stability of the whole and its components. The concentration of ferritin in the plasma depends on the amount of iron stored in the body. When Fe^{2+} ion is attached to ferritin, it is oxidized to the third oxidation state (Fe^{3+}); it is caused by the high toxicity of Fe^{2+}. The iron ion Fe^{2+} is a substrate in the Fenton reaction; between it and hydrogen peroxide (H_2O_2) ¬. As a result of this reaction, the iron ion is oxidized, the hydroxyl anion, and the hydroxyl radical (OH·) is formed – a reactive form of oxygen (ROS). The genes encoding the light-type protein subunit are located at locus 19q13.3; and the genes responsible for coding the H subunit are located on 8 chromosomes: 1, 2, 3, 6p21, 11, 14, 20 and Xq23-Xqter (Arosio et al. 2017).

The protein was discovered in 1937 by Victor Laufberger as an isolate from horse spleen; about 23% of its dry weight was ferritin. Its presence in human blood was confirmed several years later (Chakraborti et al. 2019). In 1972 there was the first quantification of the amount of ferritin together with the amount of antiferritin antibodies in the serum; this was done by Addison GM using an immunoradiometric method. In addition, the relationship between the concentration of iron in the blood and the protein of ferritin was established: people with anemia showed a reduced concentration of protein in the blood, and people with high concentration of iron in the blood also had a higher concentration of protein in the serum. This allowed for the deletion of the laboratory reference range for ferritin. Referring also to these studies, in 1975 Jacobs and Worwod suggested that the determination of serum ferritin may be an important and convenient method of assessing the state of iron storage in the human body. The concentration of the described protein in the serum has been measured to this day, however, it is known that its concentration is directly influenced by other factors, such as:

- inflammation
- infection
- tumors

The most surprising fact is that despite the long history of research on this protein, the mechanism of the relationship between its concentration and the above-mentioned factors is still not fully understood (Chakraborti et al. 2019).

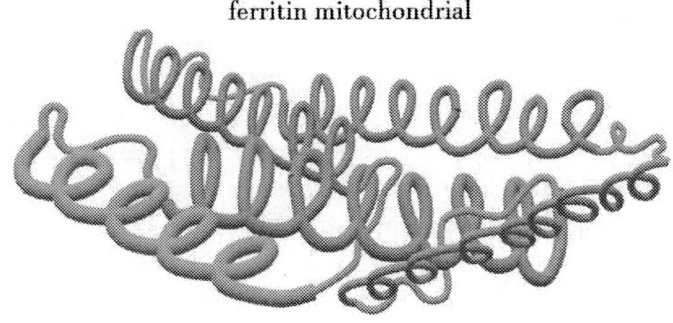

Figure 1. Ferritin Mitochondrial [Own elaboration].

In recent years, the discovery of human mitochondrial ferritin (FtMt) has been made. FtMt is a protein with a mass of approximately 30 kDa. It was described as having no intron gene on chromosome 5q23.1; is responsible for encoding a 24-amino acid precursor sequence, similar in structure to the heavy chain of the described protein (FtH). It is modulated in the mitochondria to ultimately a 22 kDa subunit that exhibits ferroxidase activity, such as FtH. This is the first-ever discovery of mammalian ferritin that targets an organelle. It is worth noting that the expression of genes encoding FtMt occurs at a much lower level than the other two ferritin chains. Particularly high expression activity for FtMt was detected in the nuclei of organs whose cells are relatively poor in mitochondria. An important basis for further research is the fact that the FeMt level does not correlate with the iron content in a given organ (Fuhrmann et al. 2020). This is noticeable by analyzing the concentration of mitochondrial ferritin in iron-rich organs such as the liver or herring. Recent clinical studies show that an increased concentration of FtMt is observed in organs showing high oxygen consumption, i.e., in cells with increased oxygen metabolism: cardiomyocytes, neurocytes, and spermatocytes. High gene expression for FtMt is observed in the cerebral cortex of patients with Alzheimer's disease and, in fact, of black substance in patients with

Parkinson's disease and restless legs syndrome. These studies indicate that mitochondrial ferritin plays a neuroprotective role by regulating apoptosis and reducing the oxidative effect of iron excess, which causes oxidative stress in the course of the Fenton reaction. Experiments in mice showed that those with a deletion of the gene sequences responsible for the synthesis of FtMt showed a lower concentration of sperm in the semen. Currently, it is postulated that the consequence of increased iron concentration in a given organ is increased FtMt synthesis, which protects cells against oxidative damage (Yévenes et al. 2017).

Intracellular ferritin is degraded in the course of two mechanisms: lysosomal and proteasomal. Under conditions of reduced serum iron concentration, nuclear coactivator 4 (NCOA4) transports ferritin to the lysosome in order to release iron ions from it, for which the body has a given requirement. This process highlights the role of ferritinophagy as a mechanism influencing the maintenance of iron cell homeostasis. NCOA4 only has an affinity for the heavy chain, not binding to FtL. Under conditions of excess iron in the blood, NCOA4 is ubiquitinated by a ubiquitin ligase (HERC2), leading to its degeneration (Fuhrmann et al. 2020).

Ferritin is a protein necessary for the proper functioning of the entire body. Scientific research on ferritin, as well as other proteins, shows perfectly well how many unknowns science still has with it. In particular, much attention is paid to the role of mitochondrial ferritin. Studies in mice clearly show that those with the FtMt gene present in cardiomyocytes are much better at coping with oxidative stress; the emerging ROS, oxidized lipids and proteins. In such mice, the synthesis of FtMt is upregulated. In mice with an inactivated gene, due to an increase in ROS concentration, cardiomyocyte damage occurs much faster with an increase in the concentration of the pro-inflammatory systemically acting factor Il-6 (Wang et al. 2019). Research on ferritinophagy indicates the presence of other, more intrusive, previously unknown, apoptotic mechanisms that enable the regulation and maintenance of homeostasis. A more detailed understanding of such mechanisms may be a place of action for potential chemotherapeutic agents or other active substances. The described phenomena make it possible to understand a number of processes aimed at providing iron ions for the synthesis of certain substances in the body; especially in the course of hematopoiesis. Ferritin is undoubtedly a protein that will continue to be the basis of research to determine the role of various factors in the process of maintaining iron homeostasis in the body (Wang et al. 2019).

References

Arosio P, Elia L, Poli M. Ferritin, cellular iron storage and regulation. *IUBMB Life.* 2017 Jun;69(6):414-422.

Chakraborti S, Chakrabarti P. Self-Assembly of Ferritin: Structure, Biological Function and Potential Applications in Nanotechnology. *Adv Exp Med Biol.* 2019;1174:313-329.

Fuhrmann DC, Mondorf A, Beifuß J, Jung M, Brüne B. Hypoxia inhibits ferritinophagy, increases mitochondrial ferritin, and protects from ferroptosis. *Redox Biol.* 2020;36:101670.

Wang W, Liu Z, Zhou X, Guo Z, Zhang J, Zhu P, Yao S, Zhu M. Ferritin nanoparticle-based SpyTag/SpyCatcher-enabled click vaccine for tumor immunotherapy. *Nanomedicine.* 2019;16:69-78.

Yévenes A. The Ferritin Superfamily. *Subcell Biochem.* 2017;83:75-102.

Chapter 19

Keratin

Adrianna Antoszewska, Dorota Bartusik-Aebisher[*] and David Aebisher

Medical College of The University of Rzeszów, Rzeszów, Poland

Abstract

>Intermediate keratin filaments are abundant in the epithelium and form cytoskeleton networks that are responsible for functions specific to a given cell type, such as adhesion, migration and metabolism. These functions are supported by the continuous cycle of keratin fiber replacement. This multi-step process keeps the cytoskeleton in motion, facilitating rapid and protein-biosynthetic-independent reconstruction of the network while keeping it intact. All epithelial cells have a visible network of keratin intermediate fibers (IF) in their cytoplasm. Keratins not only provide physical integrity in epithelial cells, but also play important roles in metabolic processes. The current goal to be achieved is to understand the molecular mechanisms of regulation of the keratin cycle in relation to the action network and microtubules, and in the context of the function of epithelial tissue.

Keywords: keratins, epithelial cells, intermediate fibers (if), cell membrane, hepatocytes

The term "keratin" has historically meant proteins obtained from the conversion of structures such as claws, hooves and horns. Currently, however,

---

[*] Corresponding Author's Email: dbartusikaebisher@ur.edu.pl.

In: The Biochemical Guide to Proteins
Editors: David Aebisher and Dorota Bartusik-Aebisher
ISBN: 979-8-88697-493-5
© 2023 Nova Science Publishers, Inc.

the term includes all the proteins that make up the intermediate filaments with specific physicochemical properties, synthesized in each epithelium of vertebrates. 'Keratin' is often misunderstood as a single substance, despite the fact that it consists of a complex mixture of proteins such as keratin, KFAP and enzymes extracted from the epidermis (Shavandi et al., 2017). Keratins are found only in epithelial cells and have specific physicochemical properties. They are insoluble in water, dilute acids, bases and organic solvents as well as resistance to digestion by proteases, pepsin or trypsin. Keratins are soluble in solutions containing denaturing agents such as urea. The basic structure of all keratins is a chain of amino acids, which can vary in amino acid sequence as well as in polarity, charge, and size. Nevertheless, the amino acid sequence of a given keratin is similar in different species (Shavandi et al., 2017).

The keratins in the respective cells and tissues of different mammalian species may have similar functions as well as antigenic epitomes that bind to the same antibody, but are in fact different in molecular structure and amino acid composition. Hence, these isoforms have a different molecular weight ranging from 40 to 70 kDa. The keratins in the simple, non-stratified epithelium are of a different type than in the layered epithelium. Epithelial cells, both in the simple and the stratified epithelium, always regularly produce individual keratins. These keratins are referred to as epithelial primary keratins, such as K8/K18 in squamous epithelium or K5/K14 in layered epithelium. Epithelial cells can also produce secondary keratins such as K7/19 in squamous epithelium or K15 and K6/K16 in layered epithelium (Mercer et al., 2019).

Keratins not only provide physical integrity in epithelial cells, but also play important roles in metabolic processes. In the intermediate layer of the layered epithelium, cells undergo various differentiation processes, such as keratinization. Intermediate filaments in epithelial cells build keratins, which are found only in vertebrates. These proteins account for approximately 80% of the total protein content in the differentiated cells of the stratified epithelium. The expression of the keratins that make up these filaments may change when epithelial cells transform into mesenchymal cells and vice versa (Mercer et al., 2019).

Keratins significantly affect the characteristics of epithelial cells, such as:

- polarity,
- the shape of cells,
- mitotic activity.

The most well-known function of keratins and keratin fibers is to contribute to the formation of a kind of scaffold for epithelial cells and tissues so that they can withstand mechanical stress, maintain their structural integrity, provide mechanical resistance, protect against hydrostatic pressure changes and establish cell polarity. Keratin filaments can be remodeled quickly, thus ensuring the flexibility of the cytoskeleton. Keratins, as components that make up keratin filaments, are concentrated at the cell periphery near the focal junctions of the cell membrane, and the polymerization of these filament-forming keratins is regulated by signaling molecules (e.g., heat shock proteins, regulatory proteins 14-3-3, and various kinases and phosphatases) (Polari et al., 2020). Keratins and keratin filaments are also responsible for non-mechanical functions such as cell signaling, cell transport, and cell differentiation. For example, K10 inhibits cell proliferation in suprabasal cells but induces their differentiation. Keratin filaments are also involved in the cell's response to internal and external signals, such as pro-apoptotic signals, and to the correct distribution of membrane proteins in polarized epithelial cells. Interestingly, in hepatocytes, K8 and K18 bind to signaling molecules, thereby disrupting the apoptotic signaling cascade that would initiate apoptosis (Qiu et al., 2020). The K18 rod domain can bind to the C-terminus of a membrane-bound signaling receptor, thereby blocking the activation of a second messenger, an enzyme involved in apoptosis in hepatocytes (Polari et al., 2020).

Unpolymerized keratins into heterodimers are broken down. Those that need to be broken down in hepatocytes are labeled for proteolysis in the proteasome by ubiquitin labeling.

Keratins – types I and II – are filamentous intermediate proteins found in epithelial cells. They are encoded by 54 genes (28 type I, 26 type II) and are regulated in a pair-dependent, tissue-type and differentiation-dependent manner. Genetically determined changes in keratin coding sequences underlie highly penetrating and rare disorders whose pathophysiology reflects cell fragility or altered tissue homeostasis. Intermediate filaments are one of the three main groups of cytoskeleton filaments in higher eukaryotes (Qiu et al., 2020). The term "intermediate filaments" means that their diameter, ~ 10 nm, is intermediate between actin microfilaments (~ 6-8 nm) and microtubules (~ 25 nm). Keratins (formerly "cytokeratins") represent types I and II IF genes and proteins that are mainly expressed in epithelial cells. Keratin proteins (Mr: 40-70 kDa) are the dominant IF subtype in all epithelials and, with a total of 54 conserved functional genes and proteins, represent nearly three-quarters of the entire mammalian IF superfamily (Zhang et al., 2019).

According to their properties, abundance and intracellular organization, it is no wonder that keratin filament assemblies contribute significantly to the mechanical resistance exhibited by several types of epithelial cells and tissues in vivo. Naturally arising mutations that disrupt the structure of 10-nm keratin filaments, their arrangement into intracellular networks in epithelial cells and/or their regulation, result in cell fragility or altered responses to cellular stress, and are responsible as a cause or predisposition to several diseases affecting the surface epithelium or inner and hair (Zhang et al., 2019).

Keratins perform many important cellular functions, and their most important common role is to protect cells and tissues from injury. The epithelium is exposed to many forms of stress. The primary function of keratin in vivo is to impart to cells the structural elasticity they need to withstand physiological doses of mechanical stress. Accordingly, IF networks are particularly well developed in tissues and cell types exposed to significant mechanical forces, such as muscle and the surface epithelium.

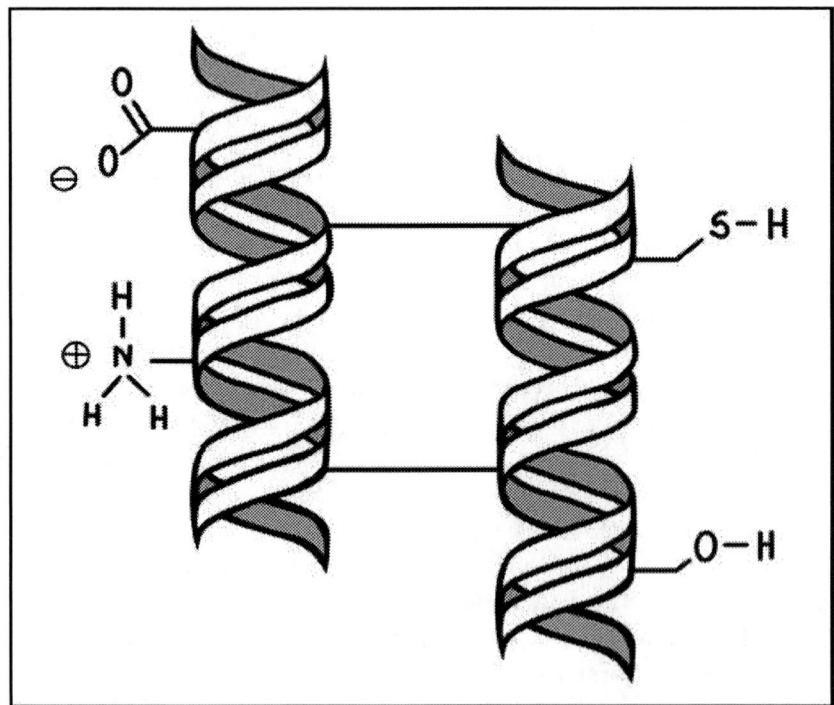

Figure 1. The structure of keratin, a source of inspiration for drawing -MDPI and ACS Style Donato, R.K.; Mija, A. Keratin Associations with Synthetic, Biosynthetic and Natural Polymers: An Extensive Review. *Polymers* 2020, *12*, 32 [Own study].

Intermediate keratin filaments are abundant in the epithelium and form cytoskeleton networks that are responsible for functions specific to a given cell type, such as adhesion, migration and metabolism. These functions are supported by the continuous cycle of keratin fiber replacement. This multi-step process keeps the cytoskeleton in motion, facilitating rapid and protein-biosynthetic-independent reconstruction of the network while keeping it intact. The current goal to be achieved is to understand the molecular mechanisms of regulation of the keratin cycle in relation to the actin network and microtubules, and in the context of the function of epithelial tissue. All epithelial cells have a visible network of keratin intermediate fibers (IF) in their cytoplasm.

References

Mercer D K, Stewart C S. Keratin hydrolysis by dermatophytes. *Med. Mycol.* 2019 Jan 1;57(1):13-22.

Polari L, Alam C M, Nyström J H, Heikkilä T, Tayyab M, Baghestani S, Toivola D M. Keratin intermediate filaments in the colon: guardians of epithelial homeostasis. *Int. J. Biochem. Cell Biol.* 2020 Dec;129:105878.

Qiu J, Wilkens C, Barrett K, Meyer A S. Microbial enzymes catalyzing keratin degradation: Classification, structure, function. *Biotechnol. Adv.* 2020 Nov 15;44:107607.

Shavandi A, Silva T H, Bekhit A A, Bekhit A E A. Keratin: dissolution, extraction and biomedical application. *Biomater Sci.* 2017 Aug 22;5(9):1699-1735.

Zhang X, Yin M, Zhang L J. Keratin 6, 16 and 17-Critical Barrier Alarmin Molecules in Skin Wounds and Psoriasis. *Cells.* 2019 Aug 1;8(8):807.

Chapter 20

GroEL

Karolina Drygała, Dorota Bartusik-Aebisher[*] and David Aebisher

Medical College of The University of Rzeszów, Rzeszów, Poland

Abstract

GroEL is one of the chaperones that play an essential role in folding non-native proteins and preventing their aggregation. GroEL is a large homotetradecamer composed of two seven-membered rings consisting of two seven-membered 57 kD subunits. It has been found that the first GroES interacting with GroEL does not always dissociate from the symmetric complex prior to dissociation of the second GroES molecule, i.e., dissociation of GroES molecules from this complex may occur in a random order. It was also found that GroEL emerged in three different states: as GroEL itself, as an asymmetric complex and as a symmetric complex. This finding points to the existence of two reaction cycles in the GroEL-GroES interaction: an asymmetric cycle and a symmetric cycle.

Keywords: GroEL, monomolecular test, non-native substrate proteins, polypeptides

GroEL is one of the chaperones that play an essential role in folding non-native proteins and preventing their aggregation. GroeEL allows folding of a single polypeptide chain to proceed unhindered by intermolecular interactions. In

---

[*] Corresponding Author's Email: dbartusikaebisher@ur.edu.pl.

In: The Biochemical Guide to Proteins
Editors: David Aebisher and Dorota Bartusik-Aebisher
ISBN: 979-8-88697-493-5
© 2023 Nova Science Publishers, Inc.

particular, GroEL performs two activities that assist the folding of proteins. One of them is the binding of coiled, partially structured semi-finished products folded in the GroEL central channel, which prevents aggregation. Such bonding may be related to a structural rearrangement which boils down to partial unfolding of non-native forms recognized by exposing their hydrophobic surface. The second activity is to facilitate the folding occurring inside the central canal after it is closed by the cofactor GroES in the presence of ATP. GroEL is a large homotetradecamer composed of two seven-membered rings consisting of two seven-membered 57 kD subunits. The GroEL subunit consists of 547 amino acids folded into three domains – the ATPase equatorial domain, the hinge intermediate domain, and the apex domain. Each domain consists of an orthogonal beta sheet structure surrounded on the inside and outside by alpha helices (Fourie et al., 2020).

GroeEL is found in the cytoplasm of eubacteria and the soluble matrix of mitochondria and chloroplasts. It was identified between 1972 and 1973 by genetic means, while looking for factors involved in bacteriophage replication, and its name reflects this story: "Gro" stands for phage growth. "E" indicates that the growth defect can be controlled by a mutation in the phage head E gene; and "L" stands for large subunit.

A widely accepted model of the GroEL-GroES reaction cycle assumes that one of the GroEL rings traps a non-native substrate protein via its hydrophobic sites, and that GroES binds to the same ring (cis ring) in an ATP-dependent manner. Binding of GroES induces displacement of the captured protein into the GroEL cavity where folding occurs. Then, as a result of ATP hydrolysis in the cis ring, ATP is bound to the opposite ring (trans ring). This dissolves the cis ring, thereby releasing GroES and the partially folded protein. At the same time, the second GroES binds to the trans ring to reorient the new cis ring and begins the next cycle of ATPase. Due to the fact that GroES binds alternatively to each GroEL ring, an asymmetric GroEL-GroES complex exists throughout the reaction cycle – also known as a ball-shaped complex in which one GroES binds to one end of a GroEL. In contrast, the symmetrical GroEL- (GroES) 2 complex, also known as the ball-shaped complex for American football, in which two GroES molecules simultaneously hood both ends of the GroEL, is not formed (Ishii et al., 2017).

In the previously studied interaction of GroEL-GroES in the reaction cycle, it was found that the course of the cycle was significantly influenced by the presence of non-native substrate proteins. In the presence of these proteins, the GroEL-GroES interaction was then monitored using Förster resonance energy transfer, without stopping the reaction. As a result, it was found that symmetric and asymmetric complexes coexist in the presence of non-native substrate proteins and that symmetric complex formation is promoted by increasing the concentration of these proteins. On the other hand, in the absence of a non-native substrate protein, a symmetric complex is not formed. This shows that the symmetric complex exists as an intermediate in the presence of a sufficient amount of non-native substrate protein. Based on an earlier report, Iizuka and Funatsu assumed that the second GroES could bind to the trans ring of the asymmetric ATP-bound complex. They then conducted similar experiments using the ATPase defective mutant (GroELD398A). GroELD398A undergoes a conformational change, as does wild-type GroEL, to bind GroES upon ATP binding, although the ATPase activity in this mutant is significantly reduced. They found that GroELD398A forms a symmetric complex when both rings are occupied by ATP. At the same time, Koike-Takeshita showed that GroELD398A forms a symmetrical complex in the presence of ATP and GroES. These results are surprising as the adopted model assumes that GroELD398A forms an asymmetric complex in the presence of ATP and GroES, and the ATP-bound complex cannot bind non-native substrate protein and GroES to the trans ring. Thus, the adopted model was questioned (Nakamoto et al., 2017).

An attempt was then made to test the GroEL-GroES interaction cycle via a symmetric complex using a monomolecular test. The test allows direct observation and characterization of symmetric complexes during the reaction cycle. It has been found that the first GroES interacting with GroEL does not always dissociate from the symmetric complex prior to dissociation of the second GroES molecule, i.e., dissociation of GroES molecules from this complex may occur in a random order. Probably, GroES dissociates with the GroEL ring, where ATP hydrolysis takes place. It was also found that GroEL emerged in three different states: as GroEL itself, as an asymmetric complex and as a symmetric complex. This finding points to the existence of two reaction cycles in the GroEL-GroES interaction: an asymmetric cycle and a symmetric cycle (Thirumalai et al., 2020).

Based on the above research, Iizuka and Funatsu proposed a new model of the GroEL-GroES reaction cycle. This model consists of two cycles: the "asymmetric cycle" and the "symmetric cycle." In the presence of a low concentration of non-native substrate protein, GroEL mainly goes through an asymmetric cycle due to the inhibitory effect of ADP on the trans ring. However, a high concentration of the non-native substrate protein causes the transition to a symmetric cycle as this protein attenuates the ADP inhibitory effect and facilitates the formation of a symmetric complex. The new model does not contradict the adopted one, but rather shows that the GroEL-GroES system can operate in a different mode in the presence of high concentrations of non-native substrate protein (Yan et al., 2018).

There are two distinct groups of chaperones: group I, found in mitochondria and chloroplasts, derived from endosymbiotic bacteria, and group II, found in archaea and in eukaryotic cytoplasm. GroEL is the best characterized chaperone.

The GroEL-GroES system is the only protein in E. coli that is essential for the growth of bacteria at all temperatures. Approximately 300 (10-15%) of the newly synthesized polypeptides isolated from E. coli cells with molecular sizes in the range of about 10-150 kDa were found to interact strongly with GroEL. Proteins larger than 60 kDa, which require the presence of GroEl for proper folding in E. coli, can be classified into two different classes. Class I requires a complete chaperone system – GroEL-GroES and ATP. In contrast, class II binds transiently to GroEL and is unlikely to require the presence of ATP and GroES.

Additionally, 85% of E. coli proteins need the help of a chaperone in one way or another during folding. Most of these polypeptides also interact post-translational with GroEL.

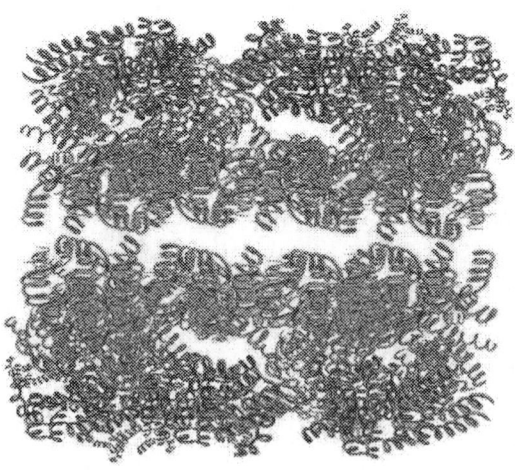

Figure 1. Structure of the GroEL-GroES complex, source of inspiration for drawing - Iizuka R, Funatsu T. Chaperonin GroEL uses asymmetric and symmetric reaction cycles in response to the concentration of non-native substrate proteins. Biophys Physicobiol. 2016 Apr 22; 13: 63-69 [Own study].

References

Fourie K. R., Wilson H. L. Understanding GroEL and DnaK Stress Response Proteins as Antigens for Bacterial Diseases. *Vaccines* (Basel). 2020 Dec 17;8(4):773.

Ishii N. GroEL and the GroEL-GroES Complex. *Subcell Biochem.* 2017;83:483-504.

Nakamoto H., Kojima K. Non-housekeeping, non-essential GroEL (chaperonin) has acquired novel structure and function beneficial under stress in cyanobacteria. *Physiol Plant.* 2017 Nov;161(3):296-310.

Thirumalai D., Lorimer G. H., Hyeon C. Iterative annealing mechanism explains the functions of the GroEL and RNA chaperones. *Protein Sci.* 2020 Feb;29(2):360-377.

Yan X., Shi Q., Bracher A., Miličić G., Singh A. K., Hartl F. U., Hayer-Hartl M. GroEL Ring Separation and Exchange in the Chaperonin Reaction. *Cell.* 2018 Jan 25;172(3):605-617.e11.

Chapter 21

Fibroblast Growth Factor (FGF)

Zuzanna Wielgosz, Dorota Bartusik-Aebisher[*] and David Aebisher
Medical College of The University of Rzeszów, Rzeszów, Poland

Abstract

Fibroblast growth factors (FGFs) were first isolated from bovine brain as protein factors with proliferative activity of 3T3 fibroblasts in in vitro cultures. FGFs are synthesized by both resident cells and migrating cells. Eating foods rich in vitamin C and amino acids allows you to increase the number of fibroblasts. Fibroblast growth factors perform a number of important functions in the body including cell proliferation, survival, migration and differentiation. Any irregularities in their functioning lead to a number of developmental defects. Future clinical applications of FGF in the regeneration of tissues, including skin, muscles, tendons, ligaments, bones, teeth and nerves, will be realized when, through appropriate use of stem cells, their biological functions are maximized.

Keywords: fibroblast growth factors (FGF), lacroacousto-dento-digital syndrome (LADD), antioxidant, mammalian cells

Fibroblast growth factors are proteins involved in many biological functions (Figure 1). They are found in organisms ranging from nematodes to humans, and have molecular weights ranging from 17 to 34 kDa and consist of at least 15 structurally related polypeptides. Among vertebrate species, FGFs are

---

[*] Corresponding Author's Email: dbartusikaebisher@ur.edu.pl.

In: The Biochemical Guide to Proteins
Editors: David Aebisher and Dorota Bartusik-Aebisher
ISBN: 979-8-88697-493-5
© 2023 Nova Science Publishers, Inc.

highly conserved both in terms of gene structure and amino acid sequence. The interaction of 23 ligands, 4 receptors and many co-receptors adds complexity to the signaling system. They take part in the transmission of various information between the cells of organisms in the embryonic period and in adulthood (Ritchie et al., 2020).

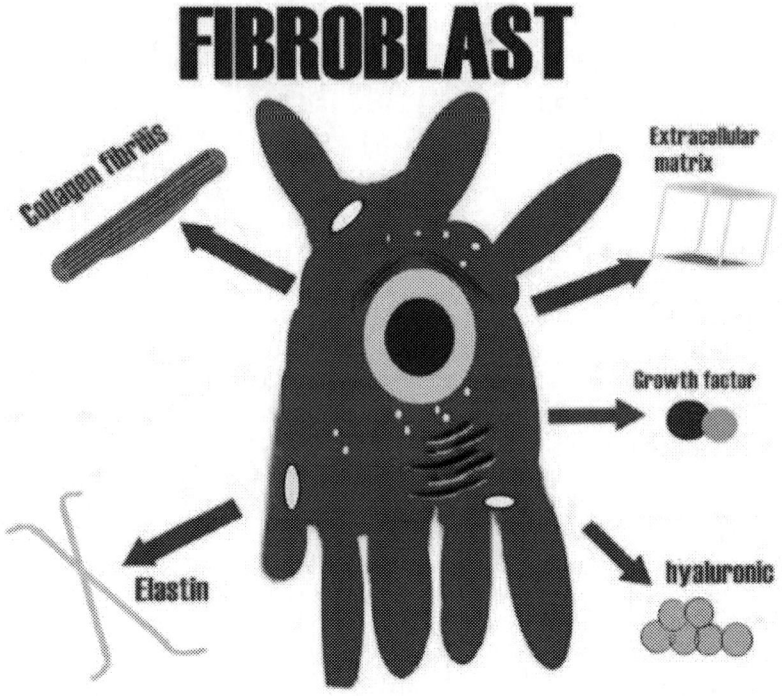

Figure 1. Influence of fibroblast growth factors on biological functions [Own elaboration].

The process of angiogenesis and wound healing are also involved. FGFs belonging to the polypeptide family bind heparin, and heparan sulfate proteoglycans are important for FGF signal transduction. The carboxyl-terminal sequence of the basic fibroblast growth factor is rich in basic amino acid residues, which is important for heparin binding. They play a useful role in the homeostasis of hematopoietic stem cells, maintaining their primitive phenotype. Primary FGF (bFGF, FGF-2) is effective in promoting the production of megakaryocytic progenitor cells during ex vivo expansion. Recent research has revealed a key role in repairing the skin, gut and liver. Minor differences in the structures and sequences of the given growth factors, their stability as well as the time of interaction with a particular type of receptor seem to be of key importance for the behavior of cells. Until then, 23 substances that belong to the group of fibroblast growth factors have been known (Li et al., 2019).

The FGF pathway plays a role in key regulatory pathways and provides a variety of potential therapeutic targets. Several germline FGF mutations have been discovered, including loss of function and enhancement mutations. The role of FGF1 is to protect the cell from stressful conditions by providing an additional signal for cell survival. This protein acts as an endothelial cell migration and proliferation modifier as well as an angiogenic factor. FGF2, also known as the primary fibroblast growth factor, induces endothelial cell proliferation and the physical organization of endothelial cells and scar-free wound healing.

The role of FGF in osteogenesis and growth was initially demonstrated by the discovery of a function enhancement mutation in FGFR3 as a cause of achondroplasia in humans. Mutations in fibroblast growth factor 3 are associated with deafness characterized by agenesis of the inner ear, microtia and microdontia. FGF4 is involved in the regulation of embryonic development, cell proliferation and differentiation. During embryogenesis, it is needed for the development of limbs and heart valves. FGF5 is required for the proper regulation of the hair cycle as it is an inhibitor of hair elongation by promoting the progression from anagen of the hair follicle growth phase to catagen. FGF6 is specific to the muscles and highly regulated during muscle regeneration. It also regulates cell proliferation, cell differentiation, angiogenesis, and myogenesis. FGF7 (called human KGF) is associated with the growth of epithelial cells. It stimulates the migration and plasminogen activity of normal human keratinocytes. Like FGF2, it plays an important role in the migration of the nerve crest cells of the midbrain

FGF8, called fibroblast growth factor, is necessary for the establishment and maintenance of the midbrain. It is involved in many processes including cell division, regulation of cell growth and maturation, and prenatal development (Ritchie et al., 2020). Fibroblast growth factor 8 by loss of function leads to Kallmann's syndrome. FGF9 signals embryonic stem cell development and sex determination. FGF9 gene expression is also essential for the development of the prostate and the maintenance of prostate homeostasis FGF10 located in the lung mesenchyma is required for the regulation of epithelial proliferation and is involved in the branching of morphogenesis in many organs such as the skin, ears and salivary glands. It is also one of the major markers of early cardiac progenitor cells and a regulator of cardiomyocyte proliferation (Li et al., 2019).

FGF11 promotes the progression of non-small cell lung cancer by regulating the hypoxia signaling pathway. FGF12 is involved in the development and functioning of the nervous system. Regulates the activity of sodium channels – supports the excitability of neurons FGF13 plays a key role in the polarization and migration of neurons in the cerebral cortex and the hippocampus, and in the regulation of sodium channels. FGF14 is a novel tumor suppressor that inhibits cell proliferation and induces cell apoptosis by mediating the PI3K/AKT/mTOR pathway FGF15 is the major hormonal FGF produced by enterocytes and is released in response to the ileal absorption of bile acids. FGF16 plays an important role in the regulation of embryonic development, cell proliferation and differentiation, and is required for normal cardiomyocyte proliferation and heart development. FGF17 is responsible for cell development in embryonic development, morphogenesis, tissue repair, tumor growth and invasion. FGF18 is involved in the regulation of cell proliferation, cell differentiation and cell migration. It is required for proper ossification and bone development, and stimulates the proliferation of the liver and intestines. FGF19 Acts as a hormone that regulates the synthesis of bile acids, influencing the metabolism of glucose and lipids. FGF20 increases the survival of cultured dopaminergic neurons in a paracrine fashion. FGF21 regulates carbohydrate and lipid metabolism and is involved in maintaining energy homeostasis and adapting to hunger and low temperature FGF22 has a protective, antioxidant and mitogenic effect on mammalian cells. High FGF22 expression is associated with lung adenocarcinoma.

FGF23 is a hormone secreted mainly by osteocytes and osteoblasts in bones. It may play a key role in mineral ion disorders. The FGF23 mutation is associated with rickets with autosomal dominant hypophosphatemia (Xie et al., 2020).

FGFs were first isolated from bovine brain as protein factors with proliferative activity of 3T3 fibroblasts in in vitro cultures. FGFs are now recognized as polypeptide growth factors and act as intracellular or extracellular signaling molecules in an endocrine, paracrine or endocrine fashion. In the case of FGF, RAS/MAP kinase is dominant. Paracrine and endocrine FGFs act through cell surface FGF receptors (FGFRs); while endocrine FGFs act independently of FGFRs. FGF signaling generally follows one of three transduction pathways: RAS/MAP kinase, PI3/AKT, or PLCγ. Each pathway regulates specific cellular behavior. Abnormal activation of pathways is associated with various pathological conditions, unregulated cell growth and neoplasm (e.g., breast and prostate cancer, lung cancer, brain and kidney tumors). Eating foods rich in vitamin C and amino acids allows you to increase the number of fibroblasts. FGFs are synthesized by both resident cells (e.g., fibroblasts) and migrating cells (e.g., macrophages). Fibroblast growth factors perform a number of important functions in the body including cell proliferation, survival, migration and differentiation. Any irregularities in their functioning lead to a number of developmental defects. Future clinical applications of FGF in the regeneration of tissues, including skin, muscles, tendons, ligaments, bones, teeth and nerves, will be realized when, through appropriate use of stem cells, their biological functions are maximized.

References

de Araújo R, Lôbo M, Trindade K, Silva D F, Pereira N. Fibroblast Growth Factors: A Controlling Mechanism of Skin Aging. *Skin Pharmacol. Physiol.* 2019;32(5):275-282.
Kuro-O M. The Klotho proteins in health and disease. *Nat. Rev. Nephrol.* 2019 Jan; 15(1): 27-44.
Li X. The FGF metabolic axis. *Front. Med.* 2019 Oct;13(5):511-530.
Ritchie M, Hanouneh I A, Noureddin M, Rolph T, Alkhouri N. Fibroblast growth factor (FGF)-21 based therapies: A magic bullet for nonalcoholic fatty liver disease (NAFLD)? *Expert Opin. Investig. Drugs.* 2020 Feb;29(2):197-204.
Xie Y, Zinkle A, Chen L, Mohammadi M. Fibroblast growth factor signalling in osteoarthritis and cartilage repair. *Nat. Rev. Rheumatol.* 2020 Oct;16(10):547-564.

Chapter 22

Hemoglobin

Szymon Płaneta, Dorota Bartusik-Aebisher* and David Aebisher
Medical College of The University of Rzeszów, Rzeszów, Poland

Abstract

Hemoglobin is a quaternary protein composed of four interconnected protein chains. In an adult human hemoglobin is most often composed of two α-globin chains and two β-globin chains. Hemoglobin is found in red blood cells and its main function is to carry oxygen (O2) molecules around the human body from the lungs to the rest of the body. A separate book can be written about this protein because all the material cannot be included in one chapter. However, there are a huge number of types of hemoglobin, many of which, whose structure has been disturbed by mutations in the genes coding for them, cause serious life-threatening diseases. The best known is sickle cell anemia, where errors in the structure of β chains cause abnormal, sickle-shaped structure of erythrocytes. This ultimately leads to anemia and can also lead to numerous congestion and death.

Keywords: hemoglobin, protoporphyrin, decarboxylation, oxygen, aminolevulinic acid (ALA)

Hemoglobin is a quaternary protein composed of four interconnected protein chains, each of which is bound to a pyrrole ring with a centrally located iron

---

* Corresponding Author's Email: dbartusikaebisher@ur.edu.pl.

In: The Biochemical Guide to Proteins
Editors: David Aebisher and Dorota Bartusik-Aebisher
ISBN: 979-8-88697-493-5
© 2023 Nova Science Publishers, Inc.

atom in the + II oxidation state. They are called heme particles. In an adult human hemoglobin is most often composed of two α-globin chains and two β-globin chains (the so-called hemoglobin H2A/HbA1). However, there are many other forms of globin that can be part of this protein, such as δ, γ, ζ, and ε. The mean weight of a hemoglobin molecule can vary, but is approximately 64.5-66.5 kDa (Ahmed et al. 2020). The genes encoding the globin α chains (HBA1 and HBA2) are located on the shorter arm of the 16th chromosome (16p.13.3). Whereas the gene encoding the β chain (HBB) is located on chromosome 11 (11p 15.15). The genes (HBE, HBG2, HBG1 and HBD) that code for the remaining forms of globin mentioned above lie in the vicinity of the β chain gene, which is called the gene cluster. Hemoglobin is found in red blood cells and its main function is to carry oxygen (O_2) molecules around the human body from the lungs to the rest of the body. It is possible thanks to the instability of binding of the above-mentioned molecules with iron atoms at the + II degree and changes in the conformation of this protein from the T form with low affinity to oxygen to the R form – with high affinity for oxygen (Ahmed et al. 2020).

As mentioned above, the most common form of hemoglobin in the human body is the form of H2A, consisting of two α-globin chains and two β-globin chains. The α chains contain 141 amino acids and the β chains contain 146 amino acids. Each of them is arranged in 7 or 8 helices, marked successively from A to H. The E and F helices form pockets for bonds. The α and β chains are connected in such a way that they form the "central water cavity" in the three-dimensional center. It is much larger in the T form of hemoglobin. The iron atom in the center of the heme porphyrin ring is stabilized by 4 nitrogen atoms and is covalently bound to the histidine imidazole ring located in the above-mentioned proximal pocket, more precisely in the F helix. This structure allows it to combine with oxygen molecules (or other gases) (Bellelli et al. 2018).

Normal amounts of hemoglobin are strictly defined. Exceeding or reducing them may indicate serious disease states. It should constitute about 34% of the mass of red blood cells (i.e., about 14-16% of the mass of blood). We call this parameter MCHC. If this value is lower than 31%, it may mean that the blood cells are undersized. Using the MCH parameter, the mean hemoglobin content of a single blood cell can be determined. It should be between 24 and 34 picograms. However, it is not as reliable as the previous one, because in the case of microcytosis (reduction in the size of red blood cells) this value will be significantly underestimated (Lim et al. 2019).

Heme biosynthesis occurs mainly in the mitochondria and can be divided into several stages. First, 8 glycine molecules condense with the same amount of active succinate that comes from the Krebs cycle. This leads to the formation of α-amino-β-ketodipic acid. Then it is decarboxylated to δ aminolevulinic acid (ALA). In the next stages, two molecules of this acid condense, thanks to which a porphobilinogen is formed, which is a porphyrin precursor. Ring synthesis occurs by the condensation of four such molecules. The resulting uroporphyrinogen is called protoporphyrin after oxidation and decarboxylation. The last step is the attachment of the Fe^{2+} ion. Globin chains (produced on the ribosomes) are attached to the 4 heme molecules thus formed, creating the final product of hemoglobin. Due to such synthesis, heme molecules can be located in the cavities of the polypeptide chain (Mandal et al. 2020).

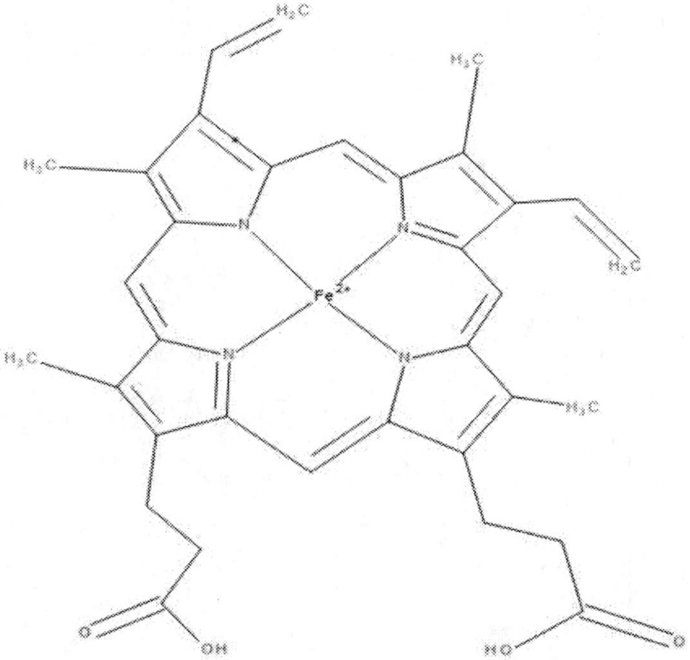

Source: Own work in the BIOVIA DRAW 2021 program.

Figure 1. Structure of a normal heme molecule.

Form T occurs when the heme molecule is not bound to oxygen. The free C-ends of all four globin chains are then not rotatable. The structure is then less mobile, more constrained, hence the abbreviation T (for tense-tense). The formation of such a structure is possible due to the numerous ionic bonds formed by the mentioned C-termini. In the R structure, all C ends can rotate, hence R (relaxated). In the T-form of hemoglobin, the iron atom in the porphyrin ring extends about 0.04 nm from the plane towards the proximal histidine. The heme group is then bulged in the same direction. During oxygenation, the iron atom, forming a strong bond with oxygen, slides into the porphyrin ring, and heme becomes flatter in relation to its plane. As it slides into the porphyrin ring, the iron atom carries the proximal histidine with it. Hemoglobin also has the ability to bind other gases, such as carbon monoxide (CO), the affinity for which is over 240 times greater than that for oxygen (O_2). It is a frequent cause of poisoning, as it turns out to be fatal above 0.1%. This is because CO binds irreversibly, unlike oxygen, which blocks the binding of oxygen and, consequently, its spreading (Zhao et al. 2021).

There are over 1,000 different forms of hemoglobin described. Most of them are incorrect forms found in various disease states. Some of these hemoglobins are listed in the table below.

Table 1. Some types of normal and abnormal hemoglobins

Type of hemoglobin	Construction	Description
HbA_1	2 chains α 2 chains β	Normal, approximately 97% of adult human hemoglobin
Hb Gower-1	2 chains ζ 2 chains ε	Occurs in fetuses, relatively rare
Hb Gower-2	2 chains α 2 chains ε	It occurs up to the third month of gestation
Hb Portland	2 chains ζ 2 chains γ	Occurs in fetuses, relatively rare
HbA_2	2 chains α 2 chains δ	Normal, approximately 2.5% of adult hemoglobin
HbF	2 chains α 2 chains γ	Occurs in fetuses and shortly after birth, it accounts for up to 0.5% in adults
HbS	2 chains α 2 chains β	Mutation in β chains, sickle cells
HbH	4 chains β	It occurs with α thalassemia

Hemoglobin is an extremely important protein with a quaternary structure. It plays an essential role in carrying oxygen around the body, which actually means sustaining life. A separate book can be written about this protein because all the material cannot be included in one chapter. The main and significant aspects of this protein have been discussed above. However, there are a huge number of types of hemoglobin, many of which, whose structure has been disturbed by mutations in the genes coding for them, cause serious life-threatening diseases. The best known is sickle cell anemia, where errors in the structure of β chains cause abnormal, sickle-shaped structure of erythrocytes. This ultimately leads to anemia and can also lead to numerous congestion and death. The disease is inherited autosomal recessively.

References

Ahmed MH, Ghatge MS, Safo MK. Hemoglobin: Structure, Function and Allostery. *Subcell Biochem.* 2020;94:345-382.

Bellelli A. Non-Allosteric Cooperativity in Hemoglobin. *Curr Protein Pept Sci.* 2018;19(6):573-588.

Lim M, Brown HM, Kind KL, Thompson JG, Dunning KR. Hemoglobin: potential roles in the oocyte and early embryo†. *Biol Reprod.* 2019 Aug 1;101(2):262-270.

Mandal AK, Mitra A, Das R. Sickle Cell Hemoglobin. *Subcell Biochem.* 2020;94:297-322.

Zhao X, Zhou J, Du G, Chen J. Recent Advances in the Microbial Synthesis of Hemoglobin. *Trends Biotechnol.* 2021 Mar;39(3):286-297.

Chapter 23

p53

Martyna Lipian, Dorota Bartusik-Aebisher[*] and David Aebisher
Medical College of The University of Rzeszów, Rzeszów, Poland

Abstract

The p53 protein is also called the genome guardian and it participates in the processes of cell aging, cell cycle arrest and apoptosis. It is activated as a result of cellular stress in the event of DNA damage, hypoxia or thermal shock. P53 is named after its molecular weight, which is 53kDa. A mutation in the TP53 gene causes the Li-Fraumeni disease syndrome. The mutation is hereditary, autosomal dominant transmission with high penetrance, and it can also arise spontaneously in the process of early embryogenesis. The p53 level in healthy cells is kept low thanks to the Mdm2 control, which is a negative regulator of p53 expression. p53 is one of the most frequently mutated genes present in cancer, up to 50%, and up to 80% in the case of very invasive forms, which leads to the conclusion that cancer therapies should focus on it. The p53 protein has been and remains in the field of interest of many research works due to the significant role it plays in inhibiting the development of neoplastic processes. Studies in which plasmids with wild-type alleles of the p53 gene were introduced into the neoplastic cells proved that in this way it is possible to achieve the effect of inhibition of the cell cycle or apopotosis. The combination of genotherapy with radio- or chemotherapy ensures increased effectiveness of the treatment.

Keywords: Li-Fraumeni disease syndrome, p53, DNA-binding domain, genotherapy

---

[*] Corresponding Author's Email: dbartusikaebisher@ur.edu.pl.

In: The Biochemical Guide to Proteins
Editors: David Aebisher and Dorota Bartusik-Aebisher
ISBN: 979-8-88697-493-5
© 2023 Nova Science Publishers, Inc.

The p53 protein is also called the genome guardian due to its suppressor and regulatory function in relation to the cell division undergoing it, it also participates in the processes of cell aging, cell cycle arrest and apoptosis. It is activated as a result of cellular stress in the event of DNA damage, hypoxia or thermal shock. The first articles about it, however, wrongly assumed its role in the formation of cancer as a factor involved in uncontrolled processes of division. P53 is named after its molecular weight, which is 53kDa. It belongs to transcription factors, i.e., proteins that bind to DNA in a specific area, where it regulates the transcription process. The p53 protein becomes active after phosphorylation. The p53 protein is encoded by the TP53 gene located on the short arm of chromosome 17, 17p13.1 to be precise (Duffy et al. 2017).

A mutation in the TP53 gene causes the Li-Fraumeni disease syndrome. The mutation is hereditary, autosomal dominant transmission with high penetrance, and it can also arise spontaneously in the process of early embryogenesis. People with it are exposed to the onset of the neoplastic process at an objectively young age – before the age of 45, however, the development of neoplastic changes is often observed already in childhood, with a tendency to develop further changes in the course of life. The most common types of neoplasms in connective tissue, osteosarcomas, brain tumors, breast and adrenal cortex tumors (Kanapathipillai et al. 2018).

The p53 protein was discovered in 1979 in the course of research carried out on mice, which were artificially immune to various types of cancer, and then the antigens present on the surface of their cells were examined. Its main goal was to deepen the knowledge of antigens that are specific to neoplastic cells and that were not observed in "normal" cells. Gel electrophoresis isolated, among other things, a protein with a molecular weight of 53,000 kDa, which was called p53. This protein was present on the surface of each of the examined neoplastic cells, and at the same time, it was not detected in the membranes of any healthy cells. This led to the supposition that it belongs to oncogenes, and this theory persisted until 1984, when it was noticed that the presence of the wild-type p53 allele would be of key importance for the inhibition of the neoplastic phenotype. The development of a tumor with a deletion of the short arm of chromosome 17, combined with a mutation in the same region on the second chromosome, was a clue to classify p53 as a tumor suppressor rather than an oncogene, which would require one to start uncontrolled divisions. Gene mutation (Liu et al. 2019).

The p53 level in healthy cells is kept low thanks to the Mdm2 control (murine double minute 2), which is a negative regulator of p53 expression. In the case of too much p53 protein in the cell, it binds to the N-terminus of the p53 domain and acts as an E3 ligase to degrade it in the proteasome in the ubiquitin-dependent pathway, regulating the amount of p53 in the cell's environment by negative feedback. The degradation activity of Mdm2 may be arrested in response to cellular stress, which results in phosphorylation of the amino and carboxyl terminals. Damage to the DNA helix or a serious error in the replication forks results in the activation of ATM and ATR kinases, which phosphorylate CHK2 and CHK1 kinase, respectively, which subsequently phosphorylate p53. As a result, it is impossible for the N-terminus TAD domain of p53 to fuse with Mdm2 and Mdm4 that would block p53 functions. This allows p53 to begin functioning as a transcription factor for genes later involved in the cycle of changes leading to cell cycle arrest, programmed cell death or metabolic regulation (Nakayama et al. 2019).

The p53 protein consists of 393 amino acids that make up the domains:

- Transactivation Domain 1 1-42;
- Transactivation Domain 2 43-62;
- Proline-rich region 63-101;
- DNA-binding domain 102-292;
- Nuclear Localization Signal 293-319;
- Tetramerization domain 320-355;
- Domain regulators 356-393. [Figure1.]

The majority (approximately 80%) of the mutations in the TP53 gene are missense mutations within the DNA-binding domain, and therefore the ability to attach to DNA is disrupted. In maintaining the function of p53, it is important for it to create a tetramer form, i.e., a functional form whose formation is disturbed in the case of mutations leading to a change in the shape of proteins. A specific mutation that occurs in the gamete may result in the Li Fraumeni syndrome, which is characterized by a high proportion of cancers that develop at an early age, often already in infancy. They can be caused by over 250 different mutations in the gene encoding the DNA binding domain of p53 protein. It is also possible the presence of Li Fraumeni syndrome in the absence of a mutation in the TP53 gene. The symptoms are then caused by a mutation in the hCHK2 gene, which is responsible for phosphorylating p53 and activating it. In the event of a mutation, this function is abolished, resulting in the occurrence of high-grade tumors (Sabapathy et al. 2019).

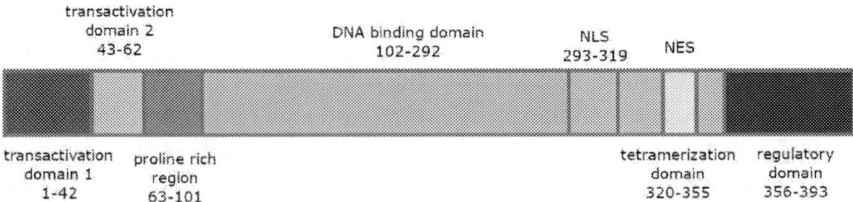

Figure 1. Distribution of p53 domains in the TP53 gene [Own elaboration].

The p53 protein has been and remains in the field of interest of many research works due to the significant role it plays in inhibiting the development of neoplastic processes. It is one of the most frequently mutated genes present in cancer, up to 50%, and up to 80% in the case of very invasive forms, which leads to the conclusion that cancer therapies should focus on it. Especially since TP53 mutations are a constant factor in many cancers, which are usually characterized by diversity both in nature and in mutation frequency for which a common source cannot be identified. There is also a pattern in mutation of the p53 gene and there are tumors with DNA sites that are more likely to mutate than others. Out of 190 mutations that can occur in the protein binding domain, only 10 lead to the occurrence of as much as 30% of missense mutations. All of the 50 most common mutations occur in the protein binding domain.

The development of a therapy aimed at resuming the normal expression, regulation and function of p53 has the potential to be successful in reducing the effects of cancer and controlling cell division. Studies in which plasmids with wild-type alleles of the p53 gene were introduced into the neoplastic cells proved that in this way it is possible to achieve the effect of inhibition of the cell cycle or apoptosis. The combination of genotherapy with radio- or chemotherapy ensures increased effectiveness of the treatment.

References

Duffy MJ, Synnott NC, Crown J. Mutant p53 as a target for cancer treatment. *Eur J Cancer.* 2017 Sep;83:258-265.

Kanapathipillai M. Treating p53 Mutant Aggregation-Associated Cancer. *Cancers* (Basel). 2018 May 23;10(6):154.

Liu J, Zhang C, Hu W, Feng Z. Tumor suppressor p53 and metabolism. *J Mol Cell Biol.* 2019 Apr 1;11(4):284-292.

Nakayama M, Oshima M. Mutant p53 in colon cancer. *J Mol Cell Biol*. 2019 Apr 1;11(4):267-276.

Sabapathy K, Lane DP. Understanding p53 functions through p53 antibodies. *J Mol Cell Biol*. 2019 Apr 1;11(4):317-329.

Chapter 24

Chitinase

Mateusz Pomianek, Dorota Bartusik-Aebisher* and David Aebisher

Medical College of The University of Rzeszów, Rzeszów, Poland

Abstract

Chitinases belong to the group of glycosidic hydrolases, their presence allows organisms, inter alia, to obtain nutrients from the decomposed polysaccharide. Chitinases, which are hydrolytic enzymes, are widely distributed in the natural environment. Their role in the defense mechanisms of individual species and in obtaining nutrients necessary for the proper development of the organism is irreplaceable due to the high prevalence of the hydrolyzed polysaccharide, which is chitin. A full understanding of the way these enzymes work, and the optimization of the process of obtaining them, most often in the form of bacterial chitinases will significantly affect the development of agriculture, forestry, and may even bring development in medicine. The products produced in reactions involving this enzyme – chitooligosaccharides, have anti-cancer and anti-inflammatory properties. The use of these products would make it possible to use targeted therapies against, inter alia, cancer cells.

Keywords: chitinases, N-acetylglucosamines, enzymes, polysaccharide, chitinase-synthesizing bacteria

---

* Corresponding Author's Email: dbartusikaebisher@ur.edu.pl.

In: The Biochemical Guide to Proteins
Editors: David Aebisher and Dorota Bartusik-Aebisher
ISBN: 979-8-88697-493-5
© 2023 Nova Science Publishers, Inc.

Chitinases belong to the group of glycosidic hydrolases, their presence allows organisms, inter alia, to obtain nutrients from the decomposed polysaccharide. They are enzymatic proteins, their presence allows the hydrolysis of β-glycosidic bonds between C1 and C4 carbon atoms of adjacent N-acetylglucosamines in the polysaccharide chitin (they act as a catalyst). These enzymes are a wide group, varying in size – the range in which these proteins fall – from 20 kDa to 120 kDa, but it should be remembered that there are differences depending on the organism from which the chitinase was isolated. Bacterial glycosidic hydrolases 20-60 kDa, plant 25-40 kDa, and enzymes from the animal kingdom (in insects) 40-85 kDa. These proteins generally work best at acidic or neutral pH, but there are also enzymes that remain active at pH values as low as 10. The broad spectrum of action of these glycosidic hydrolases can also be observed at temperature – some of them can work even at 80° C (for example chitinase derived from Streptomyces thermoviolaceus). Such versatility of the conditions in which these enzymes maintain catalytic activity allows them to be used in various processes, regardless of the conditions, while maintaining the selection of an appropriate chitinase in a given reaction environment (Du et al., 2021).

Chitinases contained in plant tissues and organs exhibit constant activity or induced by the action of an appropriate stressor. The factors regulating the activity of these enzymes can be divided into

- abiotic – the presence of heavy metals in the environment and low temperature
- biotic – protection against fungal pathogens that contain chitin, symbiosis of plants with microorganisms, protection against insects, or a method of obtaining nutrients (Itoh et al., 2019).

Two basic functions of chitinases are clearly visible, depending on their role in the body – they are signaling molecules, necessary in the symbiosis of plants with fungi and bacteria, and they also fulfill defense tasks through interaction with fungi and insects in order to eliminate them. One of the examples where chitinases are a factor necessary for the symbiosis between the plant and the microorganism is arbuscular mycorrhiza. It consists in the induction by arbuscular fungi (Glomeromycota) of the expression of the gene responsible for encoding class III chitinase. Presumably, the action of the enzymes thus synthesized in the roots of a symbiotic plant may be twofold. It inactivates the chitin elicitors derived from the fungal cell wall, which prevents the activation of the plant's defensive activities. The second possible hypothesis of the function of chitinases is to stimulate the germination of fungal spores (Oyeleye et al., 2018).

Plant protection against pathogens has been enriched in the course of evolution. Chitinases have also joined the group of methods such as: thickening and strengthening of the cell wall, free radicals release. By analyzing their amino acid sequence and biochemical properties, they were included in the PR-3, PR-4, PR-8, PR-11 protein families. All of these families share a common function of chitin hydrolysis. It should be noted that when chitinases are present as a defense mechanism, the level of their synthesis under physiological conditions is usually low. Only an infection with a pathogen stimulates their synthesis, and the factors responsible for the induction of the synthesis of these proteins are: jasmonic acid and abscisic acid. Recognition of the presence of chitin by the plant stimulates the activity of decomposing enzymes in the vacuoles, and the end result is limited fungal growth. The role of plant chitinases in symbiosis with microorganisms is confirmed, but the role of these hydrolytic enzymes on the part of bacteria is poorly understood. It was possible to isolate enzymes from individual strains of bacteria (including 12 strains of Rhizobium), and the conducted experiments against such fungi as *Aspergillus flavus*, *Aspergillus niger*, and *Fusarium oxysporum*, allow to conclude a protective role of chitinases in symbiosis. As bacteria and fungi share habitats, numerous mixed communities have arisen. An example of an ecosystem in which such a community occurs is soil, therefore it may be the largest source of chitinase-synthesizing bacteria due to the interactions that occur between them and fungi (Van Dyken et al., 2018).

These enzymes are not universal and do not show activity for all species that have chitin in their structure (which results from the different structure of, among others, individual cell walls), which limits the possibility of using them as antifungal agents with a broad spectrum of activity. Another factor that makes it difficult to obtain effective biofungicides is the difficulty of optimizing the activity through the selection of parameters such as pH of the reaction environment, temperature, or the presence of appropriate inducers. Hence, these proteins should be considered as anti-species enzymes. The combination of genetic recombination techniques with technology may allow to obtain more stable and effective enzymes that will be used to combat fungal diseases in arable fields (Van Dyken et al., 2018).

Chitinases, which are hydrolytic enzymes, are widely distributed in the natural environment. Their role in the defense mechanisms of individual species and in obtaining nutrients necessary for the proper development of the organism is irreplaceable due to the high prevalence of the hydrolyzed polysaccharide, which is chitin (the second most common polysaccharide, right after cellulose). A full understanding of the way these enzymes work, and the optimization of the process of obtaining them, most often in the form of bacterial chitinases, due to the relatively easy isolation of the enzyme from these microorganisms, will significantly affect the development of agriculture, forestry, and may even bring development in medicine, as the products produced in reactions involving this enzyme – chitooligosaccharides, have anti-cancer and anti-inflammatory properties. The use of these products would make it possible to use targeted therapies against, inter alia, cancer cells. The antifungal activity of chitinases allows the production of biofungicides from them, the task of which would be to limit the development of unwanted fungi on farmlands or in forestry. Enzymes also open the way for plants to obtain food by eating insects – they break down their chitinous shell. This property, after prior optimization of the production process of such agents, can be used to create bioinsecticides that do not adversely affect the natural environment. The high specificity of chitinases allows the production of environmentally friendly agents to combat a specific species in a given area, without adversely affecting other organisms that occur in the area of activity of this enzyme.

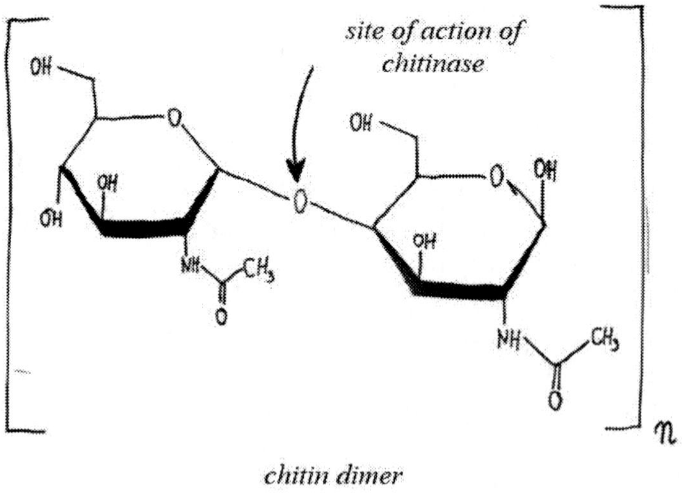

chitin dimer

Figure 1. Schematic drawing showing a fragment of the chitin chain, composed of two monomers (dimer). Visible β-glycosidic bond between the C1 and C4 atoms of two adjacent monomers. The arrow marks the site of action of chitinase-hydrolysis of this bond.

References

Du J, Duan S, Miao J, Zhai M, Cao Y. Purification and characterization of chitinase from Paenibacillus sp. *Biotechnol. Appl. Biochem*. 2021 Feb;68(1):30-40.

Itoh T, Kimoto H. Bacterial Chitinase System as a Model of Chitin Biodegradation. *Adv. Exp. Med. Biol*. 2019;1142:131-151.

Oyeleye A, Normi Y M. Chitinase: diversity, limitations, and trends in engineering for suitable applications. *Biosci. Rep*. 2018 Aug 29;38(4):BSR2018032300.

Pinteac R, Montalban X, Comabella M. Chitinases and chitinase-like proteins as biomarkers in neurologic disorders. *Neurol. Neuroimmunol. Neuroinflamm*. 2020 Dec 8;8(1):e921. doi: 10.1212/NXI.0000000000000921. PMID: 33293459; PMCID: PMC7803328.

Van Dyken S J, Locksley R M. Chitins and chitinase activity in airway diseases. *J. Allergy Clin. Immunol*. 2018 Aug;142(2):364-369.

Chapter 25

B-raf

Julia Michalik, Dorota Bartusik-Aebisher* and David Aebisher
Medical College of The University of Rzeszów, Rzeszów, Poland

Abstract

> Mutations in the human BRAF gene, which encodes the B-raf protein, may be heritable and cause birth defects, or may appear during life and contribute to the development of cancer in various organs. The B-raf protein is involved in the protein kinase signaling pathway and influences the proliferation of epithelial cells. Therefore, a mutation of this protein results in cancerous changes in tissues. Research on this reaction has led to the initiation of treatment trials for cancers caused by the B-raf mutation by utilizing the inhibition of this particular protein. However, already developed and approved combination therapies. Research is underway on new specific inhibitors of the B-raf protein. The drugs used so far show partial success. In addition to surgical treatment and radiotherapy, combined inhibitors of BRAF and MEK are introduced. The development of an effective drug that inhibits the growth of neoplastic cells resulting from these mutations would allow the fight against cancer.
>
> **Keywords:** B rapid accelerated fibrosarcoma (B-RAF), mitogen activated protein kinase (MAPK), ameloblastoma, proliferative activity

---

* Corresponding Author's Email: dbartusikaebisher@ur.edu.pl.

In: The Biochemical Guide to Proteins
Editors: David Aebisher and Dorota Bartusik-Aebisher
ISBN: 979-8-88697-493-5
© 2023 Nova Science Publishers, Inc.

BRAF is a human gene that codes for a protein called B-raf (type B rapid accelerated fibrosarcoma). It is often described in scientific literature, especially in the field of oncology. It was first mentioned in 1987, when it was isolated from the cell and classified into the raf family – the family of protein growth signal transduction kinases. It was named B-raf because it is related to, but different from, c-raf and A-raf. It is a proto-oncogenic kinase. It performs primarily regulatory functions – it is a transmitter of signals from membrane receptors for growth factors (Choi et al. 2019).

A special task of B-raf is that it belongs to the signaling pathway of kinases that are activated by mitogens (MAPK, mitogen activated protein kinase). This pathway influences cell division, differentiation and secretion. B-raf kinase is one of the messengers of this important pathway that regulates the proliferative activity of epithelial cells. In 2002, it was confirmed that B-raf occurs in a mutant form in some cancers. Its mutations have been reported in melanoma, gliomas, thyroid cancer, ovarian cancer and colorectal cancer. The B-raf kinase is made up of 766 amino acids. They form three conserved domains that are characteristic of the Raf family of kinases: conserved region 1, 2 and 3. Hydrogen bonding and other electrostatic interactions are involved in the construction of B-raf, which form dimers by binding individual kinase domains (Choi et al. 2019).

The BRAF gene, like other genes, can undergo many activating mutations. Some of them are related to the activation of the above-mentioned MAPK pathway. Against such a mutated B-raf kinase, inhibitory molecules are produced. Their development contributed to an improvement in the prognosis of patients with metastatic melanoma with a BRAF gene mutation. These observations also allowed the initiation of research into other cancers. Of these, metastatic colorectal cancer, in which mutations in the BRAF gene are frequent, can be treated with B-raf inhibition. However, a mutation of the BRAF gene in codon 600 - BRAF V600 has a particular negative impact on the prognosis. BRAF V600E is the factor that determines the sensitivity to proteasome inhibitors. This vulnerability depends on the ongoing BRAF signaling. Proteasome inhibition may be an important strategy in the treatment of colorectal cancers mediated by the BRAF V600E mutation (Köhler et al. 2019).

Mutations in the BRAF gene can cause disease in two different ways: mutations can be inherited and cause birth defects, or they can only appear later in life as an oncogene when they cause cancer. B-raf dysfunction has been observed in various neoplastic diseases. It is the most common mutant protein in melanomas. Its mutations occur with a similar frequency both in the primary tumor and in metastatic lesions of neoplasms. Most often, in 90% of cases, the mutation consists in the fact that in 1799 the thymine nucleotide is replaced with adenine. Consequently, at codon 600, valine (V) is replaced by glutamate (E), which has been called V600E (Figure 1). It is possible that the V600E mutation is the control mutation in 100% of hairy cell leukemia. It is also often found in an odontogenic neoplasm – ameloblastoma (Khan et al. 2017).

Some pharmaceutical companies are trying to develop specific mutant protein inhibitors B-raf to use them to treat cancer. In August 2011, the US Food and Drug Administration approved vemurafenib as a treatment for metastatic melanoma. This was preceded by observations and collecting the necessary clinical data. There was an improvement in patient survival and response to treatment in 53% of patients. For comparison, with the previous best available chemotherapy, dacarbazine, the response rate was approximately 12%. Unfortunately, despite the high effectiveness of the drug, 20% of tumors are still resistant to the therapy. Researchers are designing drugs to block a specific gene, BRAF, as this would help combat widespread malignant melanoma. It is estimated that about 5,000 people die each year from various cancers related to the BRAF mutation, and about 70% of melanomas are caused by such mutations (Rodriguez-Galindo et al. 2020).

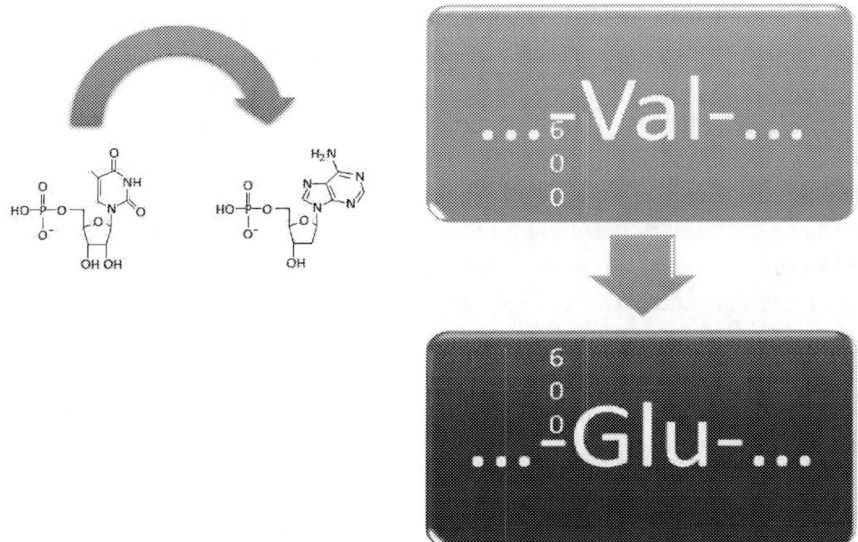

Figure 1. Mutation effect of B-raf protein; in the 1799th nucleotide adenine replaces thymine, so that in codon 600 there is glutamate instead of valine [Own elaboration].

Combined inhibition of BRAF and MEK (mitogen-activated protein kinase) was used in the treatment of melanoma. It was more effective than the BRAF inhibitor monotherapy. It is an approved treatment for metastatic melanoma caused by a BRAF mutation. Unfortunately, most patients experience disease progression even with combination therapy. Metastasis to the brain may develop. Patients undergoing therapy with a BRAF inhibitor with or without MEK inhibition should be very closely monitored clinico-radiographically for possible disease recurrence. In addition to surgery and radiotherapy, treatment with a single or dual immune checkpoint inhibitor should be initiated immediately to ensure a positive clinical outcome.

The prognosis for colorectal cancer with the B-raf mutation is extremely poor. The therapy consists in inhibiting the activity of BRAF. It is ineffective when used as monotherapy, but in combination with an anti-EGFR (epidermal growth factor receptor) drug, it has been approved for the treatment of the second metastatic lineage (Yao et al. 2019).

Mutations in the human BRAF gene, which encodes the B-raf protein, may be heritable and cause birth defects, or may appear during life and contribute to the development of cancer in various organs. They are as common in primary outbreaks as and in metastases. The most common cancer caused by such a mutation is melanoma.

The B-raf protein is involved in the protein kinase signaling pathway. Thus, it influences the proliferation of epithelial cells. Therefore, a mutation of this protein results in cancerous changes in tissues. The body produces inhibitors against the mutant kinase. Research on this reaction has led to the initiation of treatment trials for cancers caused by the B-raf mutation by utilizing the inhibition of this particular protein.

Colorectal cancer caused by a mutation of the B-raf protein, which is resistant to inhibitory monotherapy, has an extremely poor prognosis. There are, however, already developed and approved combination therapies.

Research is underway on new specific inhibitors of the B-raf protein. The drugs used so far show partial success. In addition to surgical treatment and radiotherapy, combined inhibitors of BRAF and MEK are introduced. Despite the advancement of this method, many patients exhibit resistance to this treatment and even disease progression.

Approximately 5,000 people die each year from a tumor caused by a BRAF mutation. The development of an effective drug that inhibits the growth of neoplastic cells resulting from these mutations would allow the fight against cancer associated with the B-raf protein, but would also contribute to the development of treatment of other neoplastic diseases of similar etiology.

References

Choi YW, Nam GE, Kim YH, Yoon JE, Park JH, Kim JH, Kang SY, Park TJ. Abrogation of B-Raf^{V600E} induced senescence by FoxM1 expression. *Biochem Biophys Res Commun.* 2019;516(3):866-871.

Khan MA, El-Gamal MI, Oh CH. A Progressive Review of V600E-B-RAF-Dependent Melanoma and Drugs Inhibiting It. *Mini Rev Med Chem.* 2017;17(4):351-365.

Köhler M, Ehrenfeld S, Halbach S, Lauinger M, Burk U, Reischmann N, Cheng S, Spohr C, Uhl FM, Köhler N, Ringwald K, Braun S, Peters C, Zeiser R, Reinheckel T, Brummer T. B-Raf deficiency impairs tumor initiation and progression in a murine breast cancer model. *Oncogene.* 2019;38(8):1324-1339.

Rodriguez-Galindo C, Allen CE. Langerhans cell histiocytosis. *Blood.* 202;135(16):1319-1331.

Yao Z, Gao Y, Su W, Yaeger R, Tao J, Na N, Zhang Y, Zhang C, Rymar A, Tao A, Timaul NM, Mcgriskin R, Outmezguine NA, Zhao H, Chang Q, Qeriqi B, Barbacid M, de Stanchina E, Hyman DM, Bollag G, Rosen N. RAF inhibitor PLX8394 selectively disrupts BRAF dimers and RAS-independent BRAF-mutant-driven signaling. *Nat Med.* 2019;25(2):284-291.

Chapter 26

Rhodopsin

Karolina Miś, Dorota Bartusik-Aebisher[*] and David Aebisher

Medical College of The University of Rzeszów, Rzeszów, Poland

Abstract

Rhodopsin is a biological pigment that is found in abundance in the rod cells of the retina of the eye of human and is involved in the process of vision. Rhodopsin, a receptor protein involved in the vision process, is responsible for converting the light signal into a nerve impulse. Rhodopsin is very well suited for many studies, because in the natural environment it is present in sufficient amounts that are necessary for carrying out analyzes. Rhodopsin is a GPCR receptor that is present in both humans and animals. Diseases of the retina, such as macular degeneration, are characterized by the fact that the time needed to reconstitute rhodopsin is longer than in healthy people.

Keywords: rhodopsin, helices, metarodopsin, helical chains, photoactivation

Introduction

Rhodopsin is a biological pigment that is found in abundance in the rod cells of the retina of the eye of humans and other vertebrates. The rods are active at illuminations lower than 0.1 lux and are involved in scotopic vision, i.e., with a negligible amount of light. Rhodopsin, a receptor protein involved in the

---

[*] Corresponding Author's Email: dbartusikaebisher@ur.edu.pl.

In: The Biochemical Guide to Proteins
Editors: David Aebisher and Dorota Bartusik-Aebisher
ISBN: 979-8-88697-493-5
© 2023 Nova Science Publishers, Inc.

vision process, is responsible for converting the light signal into a nerve impulse. When a single photon is absorbed, the receptor is activated, which leads to the activation (stimulation) of the G protein on the inside of the cell membrane. This is followed by the hydrolysis of cGMP to GMP by the enzyme phosphodiesterase. The decrease in cGMP concentration causes the closure of the potassium channels and depolarization of the membrane, which causes the current to flow to the synapse and, consequently, to the brain. Due to the fact that a single rhodopsin is able to activate approx. 500 G protein molecules, this signal is amplified. Rhodopsin is extremely sensitive to light, which allows you to see in low light conditions. It was discovered by Franz Christian Boll in 1876. In 1877 William Kühne named it visual purple. It is made of a protein part: opsin and 11-cis retinal, which is here a prosthetic group. 11-cis retinal is a derivative of vitamin A. Rhodopsin is a retinal chromatophore. It causes the formation of free radicals. Their generation is related to the strong metabolism of the retinal cells (Athanasiou et al., 2018).

Rhodopsin is a protein consisting of 348 amino acids, with a mass of 40 kDa. Rhodopsin is very well suited for many studies, because in the natural environment it is present in sufficient amounts that are necessary for carrying out analyzes. Structural biology allowed the study of the first crystal structure of the GPCR, rhodopsin in an inactive state. Thanks to the knowledge of the structure of rhodopsin, it was possible to develop other GPCRs structures, such as: β1 - and β2 -adrenergic receptors or the dopamine receptor (Heifetz et al., 2020).

As with other G protein-coupled receptors, rhodopsin is characterized by the motif of seven alpha-helices that pass through the lipid bilayer. They are 19AA to 34 AA long, respectively. The alpha-helices are connected to each other by six loops of different lengths. Most loops are relatively short. Hydrogen bonds, hydrophobic and ionic interactions take part in stabilizing the transmembrane helices. Helices have a different structure and are deviated from the vertical axis of the protein. Since the helices V and VI are flared on the cytoplasmic side, a Gt bond is possible. Helix VIII is the shortest, lying parallel to the lipid bilayer on the cytoplasmic side. At the end of the H: VIII helix are long palmitic chains rooted in the cell membrane. The longest loop of rhodopsin - EL: II is located between the TMH: IV and TMH: V helices, most likely it counteracts the self-activation of rhodopsin by thoroughly covering the retinal binding site. It is very important to the vision process (Meng et al., 2020).

Rhodopsin, after photoactivation, undergoes conformational changes. Inactive, visual purple absorbs a photon and then undergoes a series of transitional states: batorodopsin, lumirodopsin, and metarodopsin I. It then transforms into metarodopsin II (Meta II), which is biochemically active. Specialized methods allowed the study of rhodopsin transition structures. Their use also made it possible to elucidate the structure of photoactivated rhodopsin and to learn about its optical absorption spectrum and Gt activation ability, which turned out to be the same as in the Meta II transition state. The formation of the stable rhodopsin-Gt protein complex has critical moments. One is the stimulation of rhodopsin by light photons and the release of GDP from the nucleotide binding site in Gt. When the nucleotide remains unbound with the Gt protein – critical for the stability of the complex. Entire helixes do not move much during photoactivation, but a number of small local changes occur, especially with the ends of the V and VI helices. The above changes in the structure of rhodopsin are similar to the activation of other GPCRs caused by agonist binding, i.e., a compound which, by binding to the receptor, causes a specific reaction in the cell (Park et al., 2019).

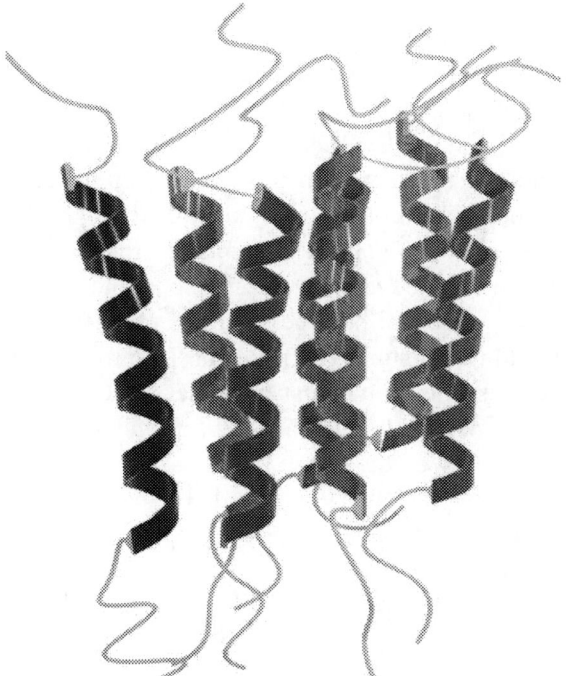

Figure 1. Model of rhodopsin structure in 3D. The figure shows seven helical chains connected by loops.

Rods adapted to the dark show very high sensitivity. Even the absorption of a single photon is sufficient to cause changes in the plasma membrane. Gt can be activated both by Meta II – it is a faster way to activate it and by opsin. However, it is the opsin that affects the "dark" noise, which is associated with increasing the sensitivity threshold. Therefore, the renewal of all visual pigments leads to the recovery of maximum sensitivity. In healthy people, restoration of rhodopsin after exposure to light occurs within 15-20 minutes. The situation is different in people who suffer from macular degeneration – in the early stages of this disease, the rods take longer to adapt to the dark. Presumably, one of the reasons may be an insufficient amount of the substrate – retinyl esters. There is increased retinal oxidative stress during AMD (Sudharsan et al., 2019).

In summary, rhodopsin is a polypeptide that is found in the rods of the eye's retina and is involved in the process of vision. The rods are very sensitive to light and allow black and white vision in low light – scotopic vision. It is made of opsin, which covalently binds to the 11 cis-retinal molecule. Absorption of the light quantum by 11-cis-retinal results in its isomerization and transformation into completely trans-retinal. Rhodopsin is a GPCR receptor that is present in both humans and animals. It was in the frog that Fraz Boll first observed it in 1876. Under the influence of light, it undergoes numerous conformational changes. Inactive rhodopsin undergoes a series of transient states and transforms into a biochemically active compound, metarodopsin II. Rhodopsin is a transmembrane protein made up of seven heliacal chains that are linked together by six loops. Individual helices differ in structure. Visual purple can be used for research because it occurs in an appropriate amount in the natural environment. Due to the knowledge of its structure, it was possible to understand other GPCRs structures. After exposure to light, it takes about 20 minutes for rhodopsin to regenerate in healthy people. Diseases of the retina, such as macular degeneration, are characterized by the fact that the time needed to reconstitute rhodopsin is longer than in healthy people.

References

Athanasiou D, Aguila M, Bellingham J, Li W, McCulley C, Reeves PJ, Cheetham ME. The molecular and cellular basis of rhodopsin retinitis pigmentosa reveals potential strategies for therapy. *Prog Retin Eye Res,* 2018; 62:1-23.

Heifetz A, Townsend-Nicholson A. Characterizing Rhodopsin-Arrestin Interactions with the Fragment Molecular Orbital (FMO) Method. *Methods Mol Biol*, 2020; 2114:177-186.

Meng D, Ragi SD, Tsang SH. Therapy in Rhodopsin-Mediated Autosomal Dominant Retinitis Pigmentosa. *Mol Ther*, 2020; 28(10):2139-2149.

Park PS. Rhodopsin Oligomerization and Aggregation. *J Membr Biol*, 2019; 252(4-5):413-423.

Sudharsan R, Beltran WA. Progress in Gene Therapy for Rhodopsin Autosomal Dominant Retinitis Pigmentosa. *Adv Exp Med Biol*, 2019; 1185:113-118.

Chapter 27

Estrogen Receptors

Julia Kudła, Dorota Bartusik-Aebisher[*] and David Aebisher

Medical College of The University of Rzeszów, Rzeszów, Poland

Abstract

Estrogens are a group of sex hormones that are derived from estran. These hormones interact by binding to specific estrogen receptors that stimulate transcriptional processes or signaling events, resulting in the control of gene expression. The estrogen receptors have many important functions during the development and maturation of tissues. Their expression is not only related to the reproductive system, but also occurs in the lungs, prostate, cardiovascular and nervous systems. In this chapter, the structural properties of the nuclear estrogen receptors Erα, Erβ of the membrane GPER1 and the mechanisms regulating gene expression through these receptors are described. The results of many studies carried out on their specification and mechanism of action have made it possible to create many therapies for various diseases in which estrogen receptors are involved in their formation. Despite the fact that these receptors were discovered in the 1960s, there are still many endocrine diseases that are associated with the abnormal interaction of estrogens, the role of their receptors has not yet been fully understood.

Keywords: estrogens, receptors, DNA, tamoxifen, clomiphene, toreomiphene, estran

---

[*] Corresponding Author's Email: dbartusikaebisher@ur.edu.pl.

In: The Biochemical Guide to Proteins
Editors: David Aebisher and Dorota Bartusik-Aebisher
ISBN: 979-8-88697-493-5
© 2023 Nova Science Publishers, Inc.

Estrogens are a group of dietary hormones that are derived from estrogen. To this group we like estrone, estradiol, estriol and estetrol. All of them are eternally called female hormones, which equally does not prevent them from appearing in the male, in which they also play an important role, e.g., in upłynnianiu nobles. Hormones release them through binding to their own estrogen receptors, which stimulate transcriptional processes or signaling events, resulting in control of gene expression. This may be due to the direct binding of estrogen receptor complexes to specific sequences in the promoters of the genes in question on genomic effects or processes that do not involve direct binding to DNA (non-genomic effects). How estrogens affect the expression of genes is regulated by complex mechanisms, regardless of whether they are hormones or secreted into the cell through the effects of nuclear or non-nuclear (Burstein et al. 2020).

Estrogen receptors, which remain in the lower part, fulfill many of the same functions during development and tissue repair. Their expression is not the only one related to the composition of the genus, but they also perform in the lungs, in the prostate, in the heart-and-nerve or nervous system. In this section, the described properties of the structural receptors of the estrogen nuclear nuclei ERα, Erβ clones GPER1 and the mechanisms regulating gene expression via emitted receptors remain. They represent the status quo history as well as their discovery and significance in the diagnosis of single units (Burstein et al. 2020).

The remaining estrogen receptors were discovered for the first time in connection with this hormone in 1958 through Elwood Jensen. He said that estrogens were extracted from the blood through the white female cells. After ten years, he discovered that estrogen-binding receptors migrate to the cell nucleus, where they affect gene expression. Since then, only one genus of estrogen-binding receptors ERα has been identified. It changed in 1996, when under the direction of Dr. Jana-Ake Gustafsson, he showed the truth of Erβ proteins in the cells of the prostate gland and in the cells of the granular ovary of the ovary. In 2012, the cloning of estrogen receptors conjugated to G GPER1 proteins was revealed by molecular cloning methods (Fuentes et al. 2019).

Among estrogen receptors, we distinguish between ER-alpha and ER-beta nuclear receptors and GPER1 bovine receptors. Nuclear receptors are encoded by two different genes, but their structure is identical, but some structural elements are distinguished by different degrees of similarity. The ER-alpha receptor is encoded by the ESR1 gene, which is located in the q24-q27 part of the sixth autosome. The Er-alpha full-length isoform has a mass of 66 kDa and consists of 595 amino acids. The second type, ie. Er-beta natomiast is encoded by the ESR2 gene on the 14q23-24 black chromosome (Figure 1). It is built of 530 ammonia, with a mass of 59 kDa. ER-beta also contains a number of isoforms (54 kDa, 49 kDa, now 44 kDa) (Hsu et al. 2017).

- the A/B domain is responsible for ligand-independent transcriptional receptor activity, but is specific to the tissue and promoter, it is bound to the transcriptional complex or to the co-regulator,
- DBD domain (DNA binding domain) its function is specific DNA binding and also dimerization of receptors
- The domain is responsible for the proper localization in the core, which can also change the conformational part.
- The LBD (ligand binding domain) is made up of 12 protein helis α, which binds the ligand binding site, which results in binding is transcription. It works with white HSP, responsible for dimmerization
- The C-terminal domain activates transcription.

Figure 1. Schematic structure of the estrogen receptor.

Estrogen is a steroid hormone due to the fact that it binds to and connects with intracellular ER-alpha and ER-beta, and is subsequently separated by binding to DNA sequences. The hormone can also trigger a cascade of reactions as a result of interaction with GPER1 or ER-beta or alpha. As a result of differences between cell reactions, which lead to the regulation of gene expression as a result of separation of the receptor-estrogen complex mediated or mediated directly by DNA, events and divided into genomic (classical) now unequal (Pakdel et al. 2018).

In the mechanism of genomic estrogen binds to the receptor, then the complex is translocated to the cell nucleus. In the knee stage, the receptor undergoes dimerization (alpha-alpha and beta-beta homodimers and alpha-beta heterodimers) and is associated with its own DNA sequence of the so-called. Estrogen response element (ERE), which is located in the promoter of delineated genes. Simultaneous estrogen intake with the receptor activates changes in the conformational receptor of the binding ligand, which makes it possible to connect coactivators. Estrogen hormones can also act without direct initiation of transcription of the whole gene and protein synthesis, which is called a non-genomic mechanism, which is weaker than the model of the classical answer (Pakdel et al. 2018).

In the inhomogeneous model, estrogen binds to the GPER1 receptor and subsequently can activate the synthesis of cAMP, a kinase of protein signaling cascades that are directly mediated by changes in gene expression, gene expression. Activation of these receptors also depends on the secretion of free hormone or other ligands in the blood, their uptake to the receptor, as well as changes released by previous activation. Estrogen opposites, which are formed through the gonads, act equally as organic and inorganic compounds that can recognize and bind to estrogen receptors. Higher selectivity is characterized by ER-alpha by ER-beta. We distinguish the main classes of estrogen receptor ligands. These are: endoestrogens, xenoestrogens, phytoestrogens, metalloestrogens, SERM. This last group of ligands is, in general, the action of those who are equally agonists and antagonists depending on the tissues. Do SERM zaliczamy tamoxifen, clomiphene, toreomiphene. Tamoxifen is one of the most commonly used drugs for the treatment of breast cancer (Tang et al. 2019).

As a result of many clinical trials, one of the risk factors for developing breast cancer is the increased exposure to estrogen. Estrogen binds to estrogen receptors, stimulates the proliferation of cells of the daily cells, intensifies cell division, DNA synthesis increases the risk of procrastination during replication, which may be related to normal, or induced apoptosis, cell proliferation or DNA production. It is also stated that the remission of the newborn after the completed treatment occurred in about 60% of patients in which the cells of the newly formed cells confirmed the expression of ER, more often than not, when the receptor was called by the receptor. It has also been shown that estrogens mediate estrogen receptors, which are related to the mentioned hormones. Badania has shown that Erα appears in the changes of modern relations of the Nublon and silver and ERbeta are the main expressions in the buttocks from grain chambers. Excess estrogen prevents osteoporosis in both areas and inhibits the activity of osteoclasts and increases osteoblasts, leading to the formation of bone tissue. The hormone is also considered to be a factor in preventing postmenopausal bone loss. In addition to the many results confirming the connection between ERalph and ERbet polymorphisms and ERB with the occurrence of osteoporosis, there is some controversy. It has also been shown that patients with a mutation in the ERalpha coding genius have incomplete plantation closures and reduced bone mineral density (Tang et al. 2019).

Estrogen receptors, which are the subject of this chapter, are involved in the regulation of numerous physiological processes. They influence cellular metabolism through genomic or non-genomic mechanisms as a result of cascades of intracellular reactions. By binding with specific ligands, they affect many different tissues in the body and mutations in the genes encoding them can cause many dysfunctions. The results of many studies on their specification and mechanism of action have allowed the creation of many therapies for various diseases in which estrogen receptors are involved. Despite the fact that these receptors were discovered in the 1960s, there are still many endocrine diseases that are associated with the abnormal interaction of estrogens, the role of their receptors has not yet been fully understood.

References

Burstein HJ. Systemic Therapy for Estrogen Receptor-Positive, HER2-Negative Breast Cancer. *N Engl J Med*. 2020;383(26):2557-2570.

Fuentes N, Silveyra P. Estrogen receptor signaling mechanisms. *Adv Protein Chem Struct Biol*. 2019;116:135-170.

Hsu LH, Chu NM, Kao SH. Estrogen, Estrogen Receptor and Lung Cancer. *Int J Mol Sci*. 2017;18(8):1713.

Pakdel F. Molecular Pathways of Estrogen Receptor Action. *Int J Mol Sci*. 2018;19(9):2591.

Tang ZR, Zhang R, Lian ZX, Deng SL, Yu K. Estrogen-Receptor Expression and Function in Female Reproductive Disease. *Cells*. 2019;8(10):1123.

Chapter 28

Ceruloplasmin

Natalia Magierło, Dorota Bartusik-Aebisher* and David Aebisher
Medical College of The University of Rzeszów, Rzeszów, Poland

Abstract

Ceruloplasmin is an acute phase protein synthesized in the liver, enterocytes and placental cells and belongs to the 2-globulin. It contains amino acids, prosthetic carbohydrates and copper. Its molecular weight is 132kDa. Disorders in its structure or quantity significantly affect health, causing various types of diseases (Wilson's disease, aceruloplasminemia, etc.) therefore it is also of great diagnostic and clinical importance. Ceruloplasmin is an example of a "lunar" protein that overcomes the concept of one gene – one structure – one function to follow changes in an organism under its physiological and pathological conditions. Due to the use of certain characteristic properties of Cp, it is possible to isolate it from blood plasma. There is a lot of interest in this protein, and a lot of medical and other research is being done on it – there are plenty of articles on the internet on this topic.

Keywords: aceruloplasminemia, ceruloplasmin, ferroxidase, Wilson's disease

Ceruloplasmin (EC 1.16.3.1.; e.g., ferroxidase) is an acute phase protein synthesized in the liver, enterocytes and placental cells and belongs to the α2-globulin. It contains amino acids, prosthetic carbohydrates and copper (usually

---

* Corresponding Author's Email: dbartusikaebisher@ur.edu.pl.

In: The Biochemical Guide to Proteins
Editors: David Aebisher and Dorota Bartusik-Aebisher
ISBN: 979-8-88697-493-5
© 2023 Nova Science Publishers, Inc.

contains six to seven copper atoms). Its molecular weight is 132kDa. It was first described in 1948 by Holmberg and Laurell. Its physiological substrates in the body are ascorbic acid, adrenaline and noradrenaline. The optimum pH for the activity of this enzyme is 5.5-5.8. The plasma of a healthy adult human contains about 115 μg Cu^{2+}/100 ml. About 95% of this copper is bound to this distinctive blue protein. Ferroxidase belongs to the family of poly-copper oxidases (the main oxidase found in plasma), which are a group of evolutionarily conserved proteins that use copper to combine substrate oxidation with the four-electron reduction of oxygen to water. Although copper is essential for the function of ceruloplasmin, this protein does not play a significant role in the transport or metabolism of this metal (Bellos et al. 2018).

It is an antioxidant and also supports iron binding with transfer, thanks to the oxidation of iron from Fe^{2+} to Fe^{3+}. In addition, it has the ability to activate the processes of ascorbic acid, norepinephrine, sulfhydryl compounds, norepinephrine oxidation and participates in the defense mechanism by participating in oxidative stress. Ceruloplasmin is an example of a "lunar" protein that overcomes the concept of one gene - one structure - one function to follow changes in an organism under its physiological and pathological conditions. Due to the use of certain characteristic properties of Cp, it is possible to isolate it from blood plasma. There is a lot of interest in this protein, and a lot of medical and other research is being done on it – there are plenty of articles on the internet on this topic. The determination of ceruloplasmin in the blood is used, inter alia, in the diagnosis of copper metabolism and Wilson's disease (Das et al. 2018).

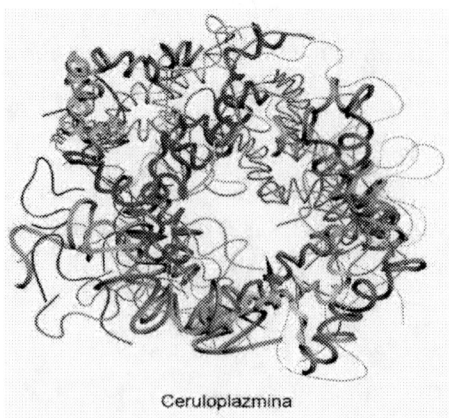

Figure 1. 3D graphic drawing of ceruloplasmin.

In the mid-twentieth century, there was a lot of research on the structure of ceruloplasmin. With the advancement of technology and methodology in 1972, Ryden's first hypothesis of a monomeric, single-chain Cp structure appeared, which was confirmed by Takahashi in 1984. Human ceruloplasmin is defined as a single-stranded protein composed of 1046 amino acid subunits divided into three units. Based on the defined amino acid sequence and taking into account the possibility of binding four glycosamine oligosaccharides, its molecular weight was determined to be 132kDa. Human Cp is very sensitive to the proteolytic activity of enzymes. The units obtained by proteolytic activation are divided according to their molecular weights into 67 kDa, 50 kDa and 19kDa. The 67 kDa fragment is designated as the N-terminus of proteins composed of 480 amino acid subunits with three glycosamine oligosaccharide binding sites (Das et al. 2018).

- Ceruloplasmin oxidizes ions Fe^{2+} do Fe^{3+} and reduces ions Cu^{2+} in protein. This reaction provides control over the formation of reactive oxygen species and the release of iron ions into the blood plasma, where it is incorporated into the apotransferrin. This control allows the unimpeded transport of iron ions in the body
- Cp can also act as an antioxidant, preventing the synthesis of free radicals during the oxidation of iron ions
- It is also a copper transport protein. About 95% of serum Cu is bound to ceruloplasmin, which carries it from the blood to other organs and tissues
- Ceruloplasmin, as a polymouth oxidase, catalyzes substrate oxidation reactions while reducing molecular oxygen to water
- Ceruloplasmin also has a bactericidal effect. Its activity as a ferroxidase reduces the availability of free iron, which in turn reduces the growth of bacteria (Lopez et al. 2021).

Ceruloplasmin belongs to the acute phase proteins. This means that its synthesis increases when homeostasis is disturbed, caused by acute inflammation, bacterial infections, cancer or autoimmune diseases. Acute phase proteins limit the spread of the inflammatory process and remove its effects.

The normal concentration of ceruloplasmin in human blood is 1.5–3.0 µmol/l (200–400 mg/l). Marking it is used, among others, in for the diagnosis of copper metabolism. We observe an increased concentration of ceruloplasmin during the occurrence of inflammations, infectious diseases, rheumatic diseases and cancer, Wilson's disease, and also during pregnancy. The deficiency of this protein may be genetically determined or result from copper deficiency. Hypoceruloplasminemia occurs when a genetically determined ceruloplasmin deficiency leads to a decrease in the serum protein level to about 50% of normal values (no clinical symptoms of this disorder). In the case when mutations completely cancel ceruplasmin activity (aceruloplasminemia), clinical symptoms are very serious (Wang Paz et al. 2019).

Aceruloplasminemia is caused by a mutation in the ceruloplasmin (CP) gene and is inherited in an autosomal recessive manner. The CP gene contains the instructions for the production of the enzyme ceruloplasmin. This enzyme is essential for the proper functioning and transport of iron in the body. Mutations in the CP gene result in a deficiency of functional ceruloplasmin, which ultimately leads to an accumulation of iron in the brain and other organs of the body. The accumulation of iron damages the tissues of the affected organs, causing the characteristic symptoms – CNS damage causes dementias, dysarthria and dystonias. Aceruloplasminemia-related insulin-dependent diabetes mellitus is the result of an accumulation of iron in the pancreas. Iron can also build up elsewhere in the body, such as the retina or the liver (Zanardi et al. 2021).

A significant decrease in serum ceruloplasmin is a biomarker of recessively inherited Wilson's disease. It is associated with a mutation in the ATPase gene that transports copper P-type (ATP7B protein), which prevents the secretion of excess copper in the bile, which leads to the accumulation of copper in the liver, brain, kidneys and red blood cells. If this disease is not treated, it causes haemolytic anemia, chronic liver disease (such as cirrhosis or inflammation), and neurological symptoms associated with copper deposition in the brain. Treatment is based on limiting the amount of copper in the food consumed and removing excess copper with penicillamine (Zanardi et al. 2021).

Ceruloplasmin is the major blood plasma oxidase that belongs to the $\alpha 2$-globulins. Due to many very important functions it performs, i.e., oxidation of Fe ions, transport of Cu ions or the activity preventing the formation of free radicals, it is necessary for the proper functioning of the body. Disorders in its structure or quantity significantly affect health, causing various types of diseases (Wilson's disease, aceruloplasminemia, etc.) therefore it is also of great diagnostic and clinical importance. People deficient in active ceruloplasmin in the peripheral blood are unable to recover Fe^{2+}, which leads to the accumulation of iron in the liver and other tissues, causing the development of insulin-dependent diabetes mellitus and damage to the central nervous system with symptoms such as dementia, dysarthria and dystonia. Cp is also an activator of some processes (e.g., ascorbic acid).

References

Bellos, I., Papantoniou, N., and Pergialiotis, V. (2018). Serum ceruloplasmin levels inpreeclampsia: a meta-analysis. *J Matern Fetal Neonatal Med.*, Sep, 31(17), 2342-2348.
Das, S., and Sahoo, P. K. (2018). Ceruloplasmin, a moonlighting protein in fish. *Fish Shellfish Immunol.*, Nov, 82, 460-468.
Lopez, M. J., Royer, A., and Shah, N. J. (2021). Biochemistry, Ceruloplasmin. 2021 May 3. In: *StatPearls* [Internet]. Treasure Island (FL): StatPearls Publishing; Jan–. PMID, 32119309.
Wang, B., and Wang, X. P. (2019). Does Ceruloplasmin Defend Against Neurodegenerative Diseases? *Curr Neuropharmacol.*, 17(6), 539-549.
Zanardi, A., and Alessio, M. (2021). Ceruloplasmin Deamidation in Neurodegeneration: From Loss to Gain of Function. *Int J Mol Sci.*, Jan 11, 22(2), 663.

Chapter 29

Prostate-Specific Antigen (PSA)

Halszka Wajdowicz, Dorota Bartusik-Aebisher[*] and David Aebisher
Medical College of The University of Rzeszów, Rzeszów, Poland

Abstract

> PSA is certainly one of the most important clinical human proteins that the scientific world has been interested in for decades. As a prostatic biomarker, it allows the detection of a large number of prostate pathologies at an early stage of development thus significantly increasing the chances of recovery or survival of the patient. It should also be mentioned that thanks to the easy and cheap form of the PSA test, it also made it possible to popularize prophylaxis in men. The deepening of the knowledge of the scientific world in this field also influenced the growing awareness of the need for research among men – this is confirmed by the annual, nationwide campaign called "MOVEMBER", which aims to make the society aware of the widespread problem of prostate cancer.

Keywords: prostate-specific antigen (PSA), peptides, N-terminal amino acids, biomarker, prostate cancer

Before considering the prostate-specific antigen, abbreviated as PSA (prostate-specific antigen), the most important information regarding this glycoprotein should be presented. It was discovered in 1970 in male semen

---

[*] Corresponding Author's Email: dbartusikaebisher@ur.edu.pl.

In: The Biochemical Guide to Proteins
Editors: David Aebisher and Dorota Bartusik-Aebisher
ISBN: 979-8-88697-493-5
© 2023 Nova Science Publishers, Inc.

and was originally called gamma semoprotein. It is the human body's best known serine protease and, as its name suggests, is produced in the prostate gland. It belongs to the tissue kallikrein family. Its presence can be found mainly in the secretory cells of the prostate ducts and in semen, where it performs its proper function – it hydrolyzes semenogelins - I and II. It would seem that this relationship is typical only for men, but the presence of PSA has been found in women, and more precisely in epithelial cells of the periurethral glands (Nordström et al. 2018). It is also worth mentioning that, like other proteins capable of hydrolyzing compounds in the human body, the prostate specific antigen is also secreted into the ducts in an inactive form as the 244-amino acid proenzyme (proPSA). The form of the described glycoprotein in the blood is also not obvious. When PSA enters the bloodstream, it binds to a specific inhibitor. However, it cannot be generalized that this is the case with every "dose" of antigen. Some are inactivated by the action of proteolysis and circulate as free PSA. It is also worth emphasizing at the very beginning how important it has become to know PSA in diagnostics – it is a biomarker of prostate cancer, hence the possibility of determining its level in the serum is so important (Nordström et al. 2018).

The human prostate antigen is an extremely important biomarker that is actively used in the clinic. Its invaluable and widespread use has contributed to the meticulous study of this glycoprotein and the great interest of the scientific world, so we can discuss it in detail. PSA is a member of tissue kallikreins, which are serine proteases that are designed to hydrolyze specific proteins with high molecular weight. This results in the formation of bioactive peptides called kinins (Duffy et al. 2020). Kallikreins can be divided into tissue and plasma levels, PSA belongs to the first group. The gene encoding the structure of PSA is present on the 19q13.4 chromosome and its expression is positively regulated by the androgen receptor which is abbreviated as AR. It is important to mention that the human prostate antigen gene is recognized as a model example of androgen controlled regulation. Interestingly, the region that accounts for high PSA levels is not located on the chromosome at the same site where transcription begins – it is upstream and is called the distal PSA enhancer. AR, and more precisely its antagonist, performs an important function of the so-called androgen blockade consisting in the use of an AR antagonist and castration, which causes a significant decrease in PSA (Duffy et al. 2020). Despite the aforementioned methods, studies have shown that they do not have a significant impact on the survival of patients. Looking at the microscopic structure of the prostate – the main organ that produces PSA, it can be concluded that it is an organ with a fairly strict structure that is extremely important – any disturbances in the architecture of the prostate caused, among others, by pathological conditions such as cancer, cause PSA to penetrate the endothelium to a greater extent into the peripheral blood, therefore an elevated level should alert the physician to a possibly developing disease. In a properly functioning prostate, seminal fluid – which is a reservoir of PSA – enters the urethra through 12-20 excretory ducts (Moradi et al. 2019).

Human prostate antigen circulates in the blood as an inactive form and its activation is dependent on the enzyme hK2, which is also mainly expressed in the prostate. Interestingly, trypsin also has the ability to activate proPSA. The action of hK2 and trypsin is based on the cleavage of the proenzyme between the seven N-terminal amino acids, and exactly the cleavage takes place between arginine (position 7) and isoleucine (position 8), which generates a mature, fully functional enzyme. Research shows that about 30% of the PSA in the plasma is a proteolytically undamaged enzyme, i.e., an inactive form, and about 5% is combined with a protein C inhibitor. Although a small part of the enzyme enters the bloodstream directly, it is an important marker of the condition of the prostate in men. It is worth mentioning that despite the undisputed organ specificity of PSA, it should be borne in mind that it does not allow to directly identify the disease entity (Pérez-Ibave et al. 2018). Increased PSA in the serum may result from damage to the prostate cells, meaning:

- trauma of the perineum,
- transurethral resection of the gland;
- TURP,
- benign prostatic hyperplasia
- inflammation,
- malignant tumor

It is also worth noting that the increased concentration of PSA in the blood can also be manifested by mechanical action such as anal intercourse or rectal examination to examine the prostate gland. Usually, in a healthy man, its level will stabilize within 2 to 3 days. The Table 1 below presents the level of PSA and the risk of developing the pathological condition in the examined condition:

Table 1. The level of PSA

PSA level	Disease risk
2.5 ng/ml	Normal value in men aged 40-50
3.5 ng/ml	Normal value in men aged 50-60
4.5 ng/ml	Normal value in men aged 60-70 years
6.5 ng/ml	Normal value in men aged 70-80 years
10.0 ng/ml and above	Incorrect value alarming about a potential disease state in the prostate, e.g., cancerchorobowym w prostacie np. nowotworze

PSA is certainly one of the most important clinical human proteins that the scientific world has been interested in for decades. This is confirmed, for example, by the number of items related to the human prostate antigen on the PubMed website.gov (one of the best-known English-language search engines for scientific articles), which for "PSA" has over 39,500 results. You can find work published over many years from 1948 to 2021. Thanks to all the research, it has been meticulously known – from structure to specific properties. This made it possible to create a simple, inexpensive and not complicated examination, which facilitated the prevention and treatment of prostate diseases. It consists only in collecting the peripheral blood into a vacuum tube and determining the concentration of PSA. As a prostatic biomarker, it allows the detection of a large number of prostate pathologies at an early stage of development, thus significantly increasing the chances of recovery or survival of the patient. It should also be mentioned that thanks to the easy and cheap form of the PSA test, it also made it possible to popularize prophylaxis in men. Regular examinations allow for quick detection of the increased concentration of the human prostate antigen and quick response of the doctor. The deepening of the knowledge of the scientific world in this field also influenced the growing awareness of the need for research among men – this is confirmed by the annual, nationwide campaign called "MOVEMBER", which aims to make the society aware of the widespread problem of prostate cancer and to encourage men to regularly prophylaxis, which is possible thanks to PSA (Johnson et al. 2021).

References

Duffy, M. J. (2020). Biomarkers for prostate cancer: prostate-specific antigen and beyond. *Clin Chem Lab Med.*, 58(3), 326-339.

Johnson, J. A., Moser, R. P., Ellison, G. L., and Martin, D. N. (2021). Associations of Prostate- Specific Antigen (PSA) Testing in the US Population: Results from a National Cross-Sectional Survey. *J Community Health.*, 46(2), 389-398.

Moradi, A., Srinivasan, S., Clements, J., and Batra, J. (2019). Beyond the biomarker role: prostate-specific antigen (PSA) in the prostate cancer microenvironment. *Cancer Metastasis Rev.*, 38(3), 333-346.

Nordström, T., Akre, O., Aly, M., Grönberg, H., and Eklund, M. (2018). Prostate-specific antigen (PSA) density in the diagnostic algorithm of prostate cancer. *Prostate Cancer Prostatic Dis.*, 21(1), 57-63.

Pérez-Ibave, D. C., Burciaga-Flores, C. H., and Elizondo-Riojas, M. Á. (2018). Prostate-specific antigen (PSA) as a possible biomarker in non-prostatic cancer: A review. *Cancer Epidemiol.*, 54, 48-55.

Chapter 30

HER2

Kacper Rogóż, Dorota Bartusik-Aebisher[*] and David Aebisher

Medical College of The University of Rzeszów, Rzeszów, Poland

Abstract

HER2 is a receptor belonging to the 185 kDa family of epidermal growth factor (EGB) receptors. Under physiological conditions, it is responsible for the regulation of the cell cycle, proper growth and proliferation of the cell. The HER2 protein plays a significant role in the vital processes of the cells in the human body. Overexpression or amplification of the gene encoding this membrane receptor can lead to oncotic, enhanced cell proliferation and growth. Research that uses electromagnetic radiation, such as computed tomography and magnetic resonance imaging, is particularly important. Current research on trastuzumab and its combination with other cytotoxic drugs shows promising results. It is important to understand the mechanism of action of this monoclonal antibody on overexpression/amplification on HER2 receptors. Herceptin may in the future prove to be a breakthrough component of the drug that inhibits cancer cells.

Keywords: HER2, epidermal growth factor (EGF), tumor cell, vinorelbine, capecitabine, computed tomography (CT), positron emission tomography (PET)

---

[*] Corresponding Author's Email: dbartusikaebisher@ur.edu.pl.

In: The Biochemical Guide to Proteins
Editors: David Aebisher and Dorota Bartusik-Aebisher
ISBN: 979-8-88697-493-5
© 2023 Nova Science Publishers, Inc.

HER2 is a receptor belonging to the 185 kDa family of epidermal growth factor (ErbB) receptors. Its counterpart was isolated in 1981. by Shih et al., in the offspring of rats that were exposed to ethylnitrosourea during pregnancy, from neuroectodermal tumor cell lines. The oncogene has been described and named as neu by Weinberg et al., in the following years. In 1985 HER2 was discovered to be a human rat protein homogen. Hence, the term neu or erbB2 was adopted for rodents. Synonymous with HER2 are CD340, NGL, p185 and erbB2 (Kunte et al., 2020).

The human membrane receptor consists of 1,255 amino acids that make up an extracellular domain with permanently open conformation and affinity for epidermal growth factor (EGF), an intracellular domain with tyrosine kinase activity, and a transmembrane domain formed by hydrophobic amino acid sequences. Under physiological conditions, it is responsible for the regulation of the cell cycle, proper growth and proliferation of the cell. Under the influence of overexpression or amplification of the HER2 gene located on chromosome 17, the number of receptors may increase significantly, which is a highly oncogenic factor found in approximately 25-40% of breast and ovarian cancers, as well as stomach, esophagus, endometrium of the throat, lungs and bladder. Currently, targeted therapies with the use of trastuzumab and the combination of this monoclonal antibody with cytotoxic drugs are known. The mechanism of action of the drug is multifaceted (Kreutzfeldt et al., 2020).

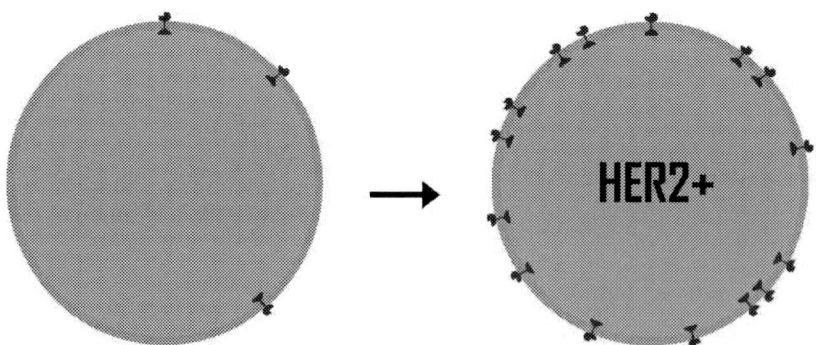

Figure 1. Comparison of a healthy cell and a cell with amplification or overexpression of the gene encoding HER2 [Own elaboration].

The action of the HER2 receptor is based on the binding of epidermal growth factor (EGF) by the extracellular domain, which triggers the transfer of phosphate residues to the 95-kDa intracellular domain, to the tyrosine and serine-threonine residues. The consequence of this is the activation of pathways related to, inter alia, MAPK and PI3K, which is associated with the formation of mTOR, which is responsible for the processes of transcription, translation, proliferation and cell growth. Overexpression or amplification of the gene encoding the HER2 protein may result in the formation of a malignant cell. In breast cancer, the expression may increase 40-100 fold and the creation of 25-50 additional copies of the HER2 gene, which results in the creation of up to 2 million receptors on the surface of the defective cell. The main mechanism of the oncogenic nature of such cells is the activation of the PI3K/Akt pathway through phosphorylation of HER3 associated with HER2. Moreover, tumor formation is exacerbated by inducing resistance of the receptor to recombinant tumor necrosis factor? (RTNF-a) and macrophages (Oh et al., 2020).

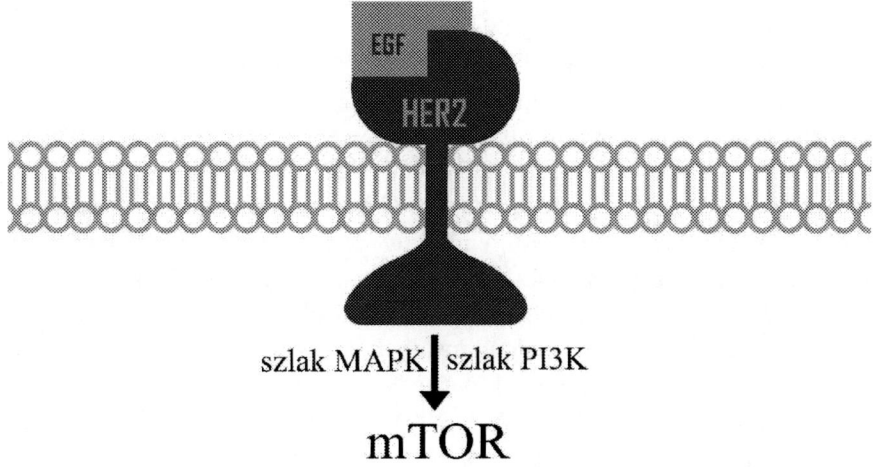

Figure 2. HER2 receptor function diagram. The attachment of epidermal growth factor (EGF) to the receptor activates intracellular pathways including MAPK and PI3K. This gives rise to the mTOR kinase, which is a direct activator of transcription, translation, proliferation and cell growth [Own elaboration].

In 1989. it was discovered that the monoclonal antibody 4D5, later named trastuzumab, could contribute to inhibiting HER2-bound cancer cells. The advantage of using monoclonal antibodies (mAbs) is their low toxicity and selective tumor targeting potential. Herceptin® (trastuzumab) is an IgG1 antibody that consists of two pairs of heavy (H) and light (L) chains with constant and variable domains with CDR1, CDR2 and CDR3 regions determining complementarity. The main mechanism of action of the drug is based on blocking the formation of heterodimers with other receptors of the ErbB family. As a result, trastuzumab indirectly inhibits the MAPK and PI3K/Akt pathways, resulting in inhibition of cell growth and proliferation. The drug also restores the p27 protein, which results in inhibition of CDK2 activity and cell cycle arrest. Another mechanism of action of herceptin is to attract NK cells to the antibody. CD16 cell differentiation antigens are involved in cellular cytotoxicity (Ocaña et al., 2020).

In addition, binding of trastuzumab to the receptor triggers the internalization and degradation of HER2 mediated by the c-Cbl ubiquitin ligase. How this down-regulation is induced and mediated remains unclear. The effectiveness of trastuzumab is 35-50% in patients given "first line" treatment and 11.6-15% in previously treated patients. The combination of herceptin with other therapeutic agents such as paclitaxel, vinorelbine, capecitabine, docetaxel, gemcitabine and platinum resulted in a 20-55% response rate. Concomitant medications bind to the receptor and then undergo endocytosis. Lysosomal breakdown of endocytic vesicles results in the release of cytotoxins which bind to the microtubules, resulting in cell death. Trastuzumab monotherapy is particularly effective in cells highly overexpressing HER2. Long-term administration of herceptin leads to secondary resistance, the cause of which has not yet been known. Theories for this mechanism speak of EGF-like ligand production, insulin-like growth factor 1 receptor overexpression, or production of the alternatively spliced extracellular erbB2 domain retained inside the cell. All possible causes share the presence of an agent of constitutive activation of the phosphatidylinositol-3-pathway, independent of erbB2. Another mechanism of resistance may be the presence of mucin, which, by forming a protective coating on the cell surface, blocks binding of Herceptin (Pernas et al., 2019).

Currently, we have many methods of detecting altered neoplastic cells. Diagnostics with high overexpression but low amplification are especially important. For ex vivo and in vitro studies, liquid chromatography, mass spectrometry, electrophoresis and immunofluorescence are used. In vivo suspicion, computed tomography (CT), positron emission tomography (PET), single photon emission computed tomography, ultrasound and optical imaging are helpful. Non-invasive proton magnetic resonance imaging is also of increasing importance.

The HER2 protein plays a significant role in the vital processes of the cells in the human body. Overexpression or amplification of the gene encoding this membrane receptor can lead to oncotic, enhanced cell proliferation and growth. A number of processes defining the neoplastic nature of such a cell, such as the MAPK, PI3K/Akt pathways, as a result of which mTOR kinase is formed, directly related to the formation of oncotic clones, can be blocked early thanks to the detection of neoplastic lesions using modern diagnostic methods. Research that uses electromagnetic radiation, such as computed tomography and magnetic resonance imaging, is particularly important. Then it is important to introduce appropriately selected treatment methods known to us. Current research on trastuzumab and its combination with other cytotoxic drugs shows promising results. For example, after a 10-year follow-up of 5099 women who participated in the 2001-2005 HERA study, the conclusion that the use of herceptin for one year significantly affects disease-free survival. The use of the drug for 2 years does not have a positive effect on the test result. It is important to conduct further research on trastuzumab and its activity when administered with other drugs. It is important to understand the mechanism of action of this monoclonal antibody on overexpression/amplification on HER2 receptors. Herceptin may in the future prove to be a breakthrough component of the drug that inhibits cancer cells.

References

Kreutzfeldt J, Rozeboom B, Dey N, De P. The trastuzumab era: current and upcoming targeted HER2+ breast cancer therapies. *Am. J. Cancer Res.* 2020;10(4):1045-1067.

Kunte S, Abraham J, Montero A J. Novel HER2-targeted therapies for HER2-positive metastatic breast cancer. *Cancer.* 2020;126(19):4278-4288.

Ocaña A, Amir E, Pandiella A. HER2 heterogeneity and resistance to anti-HER2 antibody-drug conjugates. *Breast Cancer Res.* 2020;22(1):15.

Oh D Y, Bang Y J. HER2-targeted therapies - a role beyond breast cancer. *Nat. Rev. Clin. Oncol.* 2020;17(1):33-48.

Pernas S, Tolaney S M. HER2-positive breast cancer: new therapeutic frontiers and overcoming resistance. *Ther. Adv. Med. Oncol.* 2019;11:1758835919833519.

Index

A

aceruloplasminemia, 153, 156
actin, 3, 23, 24, 25, 26, 101, 103
acute myeloid leukemia (AML), 33, 36
ameloblastoma, 135, 137
amino acid, 1, 5, 8, 18, 28, 30, 39, 40, 41, 44, 49, 50, 51, 56, 57, 60, 66, 71, 76, 87, 95, 100, 106, 111, 112, 115, 118, 125, 131, 136, 142, 149, 153, 155, 160, 166
aminolevulinic acid (ALA), 117, 119
amoebas, 23, 26
anaphylactoid reactions, 43, 44
anti-angiogenic, 75, 77, 78
antioxidant, 89, 90, 111, 114, 154, 155
autoimmunity, 81, 83

B

B rapid accelerated fibrosarcoma (B-RAF), 135, 136, 139
biomarker, 20, 156, 159, 160, 162, 163
blood plasma, 27, 30, 87, 88, 89, 153, 154, 155, 156
bone morphogenetic protein (BMP), 7
B-raf, vii, 135, 136, 137, 138, 139
bronchitis, 75, 76

C

capecitabine, 165, 168
cell division, 10, 23, 24, 26, 113, 124, 126, 136, 150
cell membrane, 8, 27, 29, 30, 51, 66, 70, 71, 82, 83, 99, 101, 142
cerebrospinal fluid (CSF), 27, 29, 30, 33, 35, 36, 49, 50
ceruloplasmin, vii, 153, 154, 155, 156, 157
chitinase(s), vii, 129, 130, 131, 132, 133
chitinase-synthesizing bacteria, 129, 131
clomiphene, 147, 150
collagen, v, 9, 17, 18, 19, 20, 21, 27, 29, 89
colony stimulating factors (CSF), 33, 35, 36
combined immune disorders (SCID), 59, 62, 63
complementarity determining regions (CDRs), 81, 83
computed tomography (CT), 165, 168, 169
C-terminal domain SGLT1, 65

D

decarboxylation, 117, 119
DNA, 7, 39, 40, 41, 42, 43, 45, 62, 123, 124, 125, 126, 147, 148, 149, 150
DNA-binding domain, 123, 125
drug-induced lupus, 39

E

elastin, v, 11, 12, 13, 14, 15, 18, 19, 27
elastogenesis, 11, 12
encephalin(s), 49, 50, 51, 52
endocytic transport, 23, 26
enzymes, 45, 100, 103, 129, 130, 131, 155
epidermal growth factor (EGF), 165, 166, 167
epidermal growth factor receptor (EGFR), 65, 67
epithelial cells, 28, 66, 99, 100, 101, 102, 103, 113, 135, 136, 138, 160
estran, 147
estrogen receptor(s), vii, 147, 148, 149, 150, 151
estrogen(s), vii, 147, 148, 149, 150, 151

Index

F

ferritin, vi, 93, 94, 95, 96, 97
ferroxidase, 94, 95, 153, 155
fibroblast growth factors (FGF), vii, 57, 111, 112, 113, 114, 115
fibronectin, v, 27, 28, 29, 30, 31
fibrosarcoma, 55, 57
fibrous protein, 17, 19
functioning, ix, 1, 5, 9, 12, 30, 35, 58, 70, 71, 87, 89, 93, 96, 111, 114, 115, 125, 156, 161
fungi, 23, 26, 130, 131, 132

G

genotherapy, 123, 126
glucocorticosteroid, 7, 9
glutamylation, 1, 3
glycosaminoglycans, 17, 20, 27, 29, 30
granulocyte-macrophage colony stimulating factor (GM-CSF), 33, 35, 37
GroEL, vii, 105, 106, 107, 108, 109

H

head and neck cancers, 65
helical chains, 141, 143
helices, 2, 17, 41, 43, 45, 46, 66, 106, 118, 141, 142, 144
hematopoiesis, 93, 96
hemoglobin, vii, 94, 117, 118, 119, 120, 121
hepatocytes, 28, 87, 88, 89, 99, 101
HER2, viii, 151, 165, 166, 167, 168, 169
histocompatibility, 69, 71, 72, 82
histones, v, 39, 40, 41, 42, 44, 46
human leukocyte antigens (HLA), 69, 70, 72
hyaluronic acid, 17, 20
hypoalbuminemia, 87, 88

I

immune system, ix, 9, 35, 45, 49, 52, 60, 62, 69, 70
immunoglobulin, 44, 59, 60, 61, 62
integrin, 28, 29, 30
intermediate fibers (if), 99, 103

K

Kawasaki disease, 59, 62, 63
keratin(s), vii, 99, 101, 102, 103

L

lacroacousto-dento-digital syndrome (LADD), 111
laminin, 27
large latent complex (LLC), 11, 13
Leśniewski-Crohn's disease, 87, 90
leu-enkephalin, 49, 50, 51
Li-Fraumeni disease syndrome, 123, 124
liver cells, 87, 89
lymphocytes, 35, 52, 60, 70, 72, 81, 82, 83, 84
lysine, 39, 40, 45

M

major histocompatibility antigens, vi, 69
major histocompatibility complex (MHC), 69, 70, 71, 72, 73, 82, 83, 84
malignant diseases, 93, 94
mammalian cells, 76, 111, 114
metarodopsin, 141, 143, 144
metastasize, 65, 67
met-enkephalin, 49, 50, 52
microblastic skin tumors, 55, 57
microtubule organization site (MTOC), 1, 4
microtubules, 1, 3, 4, 5, 99, 101, 103, 168
mitogen activated protein kinase (MAPK), 135, 136, 167, 168, 169
monomolecular test, 105, 107

multiple colony stimulating factor (MULTI-CSF), 33, 35, 37
muscle cell, 23, 24, 25, 26, 28, 55
myosin(s), v, 23, 24, 25, 26

N

N-acetylglucosamines, 129, 130
nematodes, 23, 24, 26, 111
neoplastic diseases, 7, 10, 75, 78, 81, 84, 136, 138
non-native substrate proteins, 105, 106, 108
N-terminal amino acids, 159, 161

O

oligodendrocyte progenitor cells (OPCs), 55, 57
organ transplants, 59, 62, 63
osteoarthritis, 17, 20, 115
oxygen, 55, 77, 89, 90, 94, 95, 117, 118, 119, 120, 154, 155

P

p53, vii, 123, 124, 125, 126, 127
peptides, vi, 13, 49, 50, 52, 71, 159, 160
photoactivation, 141, 142
phototransduction, 23, 24
platelet-derived growth factor (PDGF), vi, 55, 56, 57, 58
polypeptides, 105, 108, 111
polysaccharide, 129, 130, 132
positron emission tomography (PET), 165, 168
pro-angiogenic, 75, 77, 78
proliferative activity, 111, 114, 135, 136
prostate cancer, 67, 68, 115, 159, 160, 162, 163
prostate-specific antigen (PSA), vii, 159, 160, 161, 162, 163
protamines, vi, 43, 44, 45, 46, 47
proteins, ix, 1, 2, 3, 4, 5, 7, 9, 10, 12, 13, 14, 18, 23, 25, 26, 27, 29, 30, 35, 37, 39, 40, 41, 42, 43, 44, 45, 46, 60, 65, 66, 67, 69, 70, 71, 72, 78, 83, 84, 94, 96, 99, 100, 101, 105, 106, 108, 109, 111, 115, 124, 125, 130, 131, 133, 148, 154, 155, 159,160, 162
protoporphyrin, 117, 119

R

receptors, 8, 9, 10, 12, 28, 36, 37, 49, 51, 52, 56, 57, 61, 62, 78, 81, 82, 83, 84, 112, 114, 136, 142, 147, 148, 149, 150, 151, 165, 166, 167, 168, 169
rhodopsin, vii, 141, 142, 143, 144, 145

S

salmin, 43, 44, 45
Schwann cells, 27, 28
serum albumin, vi, 87, 88, 89, 91
SGLT1, vi, 65, 66, 67, 68
sulfenic acid, 87, 89
sulphates, 43, 46
supravalvular aortic stenosis (SVAS), 11, 12, 13, 15
systemic lupus erythematosus (SLE), 59, 62, 63

T

T cell receptor (TCR), vi, 81, 82, 83, 84, 85
tamoxifen, 147, 150
threonine, 45, 167
toreomiphene, 147, 150
transforming growth factors (TGFs), v, 7, 8, 9, 10, 13
tubulin, v, 1, 2, 3, 4, 5
tumor cell, 65, 66, 67, 77, 78, 83, 165, 166
tyrosination, 1, 3

U

ulcerative colitis, 87, 90

V

vascular endothelial growth factor (VEGF), vi, 75, 76, 77, 78, 79
vinorelbine, 165, 168

X

X-ray, 43, 45

About the Editors

David Aebisher, PhD, DSc
Faculty of Medicine
University of Rzeszow
Rzeszów, Poland
Email: daebisher@ur.edu.pl

Dorota Bartusik-Aebisher, PhD, DSc
Faculty of Medicine
University of Rzeszow
Rzeszów, Poland
Email: dbartusik-aebisher@ur.edu.pl